PHYLOGENETIC RELATIONSHIPS AMONG GERRHONOTINE LIZARDS

AN ANALYSIS OF EXTERNAL MORPHOLOGY

Phylogenetic Relationships Among Gerrhonotine Lizards

An Analysis of External Morphology

David A. Good

A Contribution from the Museum of Vertebrate Zoology
of the University of California at Berkeley

UNIVERSITY OF CALIFORNIA PRESS
Berkeley • Los Angeles • London

UNIVERSITY OF CALIFORNIA PUBLICATIONS IN ZOOLOGY

Editorial Board: Peter B. Moyle, James L. Patton,
Donald C. Potts, David S. Woodruff

Volume 121
Issue Date: December 1988

UNIVERSITY OF CALIFORNIA PRESS
BERKELEY AND LOS ANGELES, CALIFORNIA

UNIVERSITY OF CALIFORNIA PRESS, LTD.
LONDON, ENGLAND

ISBN 0-520-09744-0
LIBRARY OF CONGRESS CATALOG CARD NUMBER: 88-24993

© 1988 BY THE REGENTS OF THE UNIVERSITY OF CALIFORNIA
PRINTED IN THE UNITED STATES OF AMERICA

Library of Congress Cataloging-in-Publication Data

Good, David A., 1956–
    Phylogenetic relationships among gerrhonotine lizards: an analysis of external morphology / David A. Good.
    p. m. — (University of California publications in zoology; v. 121)
    "A contribution from the Museum of Vertebrate Zoology of the University of California at Berkeley."
    "December 1988"—T.p. verso.
    Bibliography: p.
    ISBN 0-520-09744-0 (alk. paper)
    1. Anguidae—Morphology. 2. Anguidae—Classification. 3. Reptiles—Morphology. 4. Reptiles—Classification.
I. University of California, Berkeley. Museum of Vertebrate Zoology. II. Title. III. Title: Gerrhonotine lizards: an analysis of external morphology. IV. Series.
QL666.L2254G66 1988
597.95—dc19   88-24993
                                              CIP

Contents

List of Figures, vii
List of Tables, ix
Acknowledgments, x

INTRODUCTION	1
MONOPHYLY OF THE GERRHONOTINAE	3
MATERIALS AND METHODS	4
CHARACTER ANALYSIS	7
PHYLOGENETIC ANALYSIS	35
HISTORY OF GENERIC TAXONOMY IN THE GERRHONOTINAE	49
KEY TO THE SPECIES OF GERRHONOTINE LIZARDS	61
SYSTEMATIC ACCOUNTS	66
Coloptychon, 66	
Gerrhonotus, 67	
Elgaria, 71	
Barisia, 78	
Mesaspis, 81	
Abronia, 87	
THE TAXONOMIC IMPLICATIONS OF GERRHONOTINE RELATIONSHIPS	102
THE BIOGEOGRAPHIC IMPLICATIONS OF GERRHONOTINE RELATIONSHIPS	103

Appendix A: Specimens Examined, 111
Appendix B: Characters Used in the Analysis of Gerrhonotine Relationships, 115
Appendix C: Distribution of Derived and Ancestral Character States Among the Genera of Gerrhonotine Lizards, 125
Literature Cited, 131

List of Figures

1. Head scale nomenclature illustrated on the proposed gerrhonotine prototype (modified from Tihen, 1949a), 6
2. Schematic representation of dorsal snout scale variation in the Gerrhonotinae, 8
3. Schematic representation of lateral snout scale variation in the Gerrhonotinae, 10
4. Schematic representation of circumocular scale variation in the Gerrhonotinae, 16
5. Schematic representation of dorsal head scale variation in the Gerrhonotinae, 17
6. Schematic representation of temporal scale variation in the Gerrhonotinae, 21
7. Photograph of a preserved specimen of *Abronia oaxacae*, 24
8. Schematic representation of chinshield variation in the Gerrhonotinae, 26
9. Photograph of a preserved specimen of *Mesaspis moreleti*, 31
10. Photograph of a preserved specimen of *Barisia imbricata*, 31
11. Photograph of a preserved specimen of *Elgaria kingii*, 31
12. Photograph of a preserved specimen of *Gerrhonotus liocephalus*, 32
13. Photograph of a preserved specimen of *Coloptychon rhombifer*, 33
14. Phylogenetic relationships among the gerrhonotine genera as suggested by the analysis of external characters, 37
15. Phylogenetic relationships among the species of *Elgaria* as suggested by the analysis of external characters, 40
16. Phylogenetic relationships among the species of *Barisia* as suggested by the analysis of external characters, 42
17. Phylogenetic relationships among the species of *Mesaspis* as suggested by the analysis of external characters, 44
18. Phylogenetic relationships among the species of *Abronia* as suggested by the analysis of external characters, 48
19. Gerrhonotine phylogeny as postulated by Smith (1942), 54
20. Gerrhonotine phylogeny as postulated by Tihen (1949a), 54
21. The phylogeny of *Barisia* (sensu Tihen, 1949a) as postulated by Tihen (1949b), 56
22. Gerrhonotine phylogeny as postulated by Stebbins (1958), 56
23. Gerrhonotine phylogeny as postulated by Waddick and Smith (1974), 58
24. Gerrhonotine phylogeny as postulated by Rieppel (1980), 58

List of Figures

25. Gerrhonotine phylogeny as postulated by Gauthier (1982), 60
26. The geographic distribution of *Coloptychon*, 68
27. The geographic distribution of *Gerrhonotus*, 70
28. The geographic distribution of *Elgaria*, 73
29. The geographic distribution of *Barisia*, 80
30. The geographic distribution of *Mesaspis*, 84
31. The geographic distribution of *Abronia*, 90
32. The major biogeographic regions based on the distribution of Middle American highland gerrhonotines, 107

List of Tables

1. Distribution of derived and ancestral states of potentially phylogenetically informative characters among the genera of gerrhonotine lizards, 36
2. Distribution of derived and ancestral states of potentially phylogenetically informative characters among the species of *Elgaria*, 39
3. Distribution of derived and ancestral states of potentially phylogentically informative characters among the species of *Barisia*, 41
4. Distribution of derived and ancestral states of potentially phylogenetically informative characters among the species of *Mesaspis*, 43
5. Distribution of derived and ancestral states of potentially phylogenetically informative characters among the species of *Abronia*, 45
6. Chronological list of species descriptions, 50

Acknowledgments

This paper represents the culmination of a substantial part of my Ph.D. dissertation work in the Department of Zoology, University of California at Berkeley. I would like to thank the faculty, staff, and students in that department, and particularly in the Museum of Vertebrate Zoology, for their support and companionship during my tenure as a graduate student. Special thanks go to the members of my dissertation committee, David Wake, Harry Greene, and Thomas Duncan, and to Aaron Bauer, Kevin de Queiroz, Jacques Gauthier, John Karges, Roy McDiarmid, James Stewart, and John Wright for important discussions and suggestions during the course of this research.

Thanks also go to the following curators for allowing me to borrow gerrhonotine material from the collections in their care (see Appendix A for museum abbreviations): P. Alberch (MCZ), N. Arnold (BMNH), J. Campbell (UTA-CV), G. Casas-Andreu (UNAM), J. Collins (ASU), R. Drewes (CAS), A. Dubois (MNHP), W. Duellman (KU), H. Greene (MVZ), R. Heyer (USNM), D. Hoffmeister (UIMNH), R. Inger (FMNH), A. Kluge (UMMZ), L. Maxson (UIMNH), R. McDiarmid (USNM), C. Myers (AMNH), R. Nussbaum (UMMZ), G. Peters (ZMB), D. Robinson (UCR), J. Wright (LACM), and R. Zweifel (AMNH). A. Knight graciously provided me with information, photographs, and illustrations of *Elgaria parva*.

Funding for this research came in part from the Museum of Vertebrate Zoology and the Zoology Department, University of California at Berkeley, and from the Natural History Museum of Los Angeles County, Los Angeles.

Some of the figures were prepared by G. Christman and others by K. Klitz.

Note: The author's present address is Museum of Vertebrate Zoology, 2593 Life Sciences Building, University of California at Berkeley, Berkeley, California 94720.

INTRODUCTION

> The alligator lizard is the species which inspires more horror in the unsophisticated mind than all the rest of our lizards put together. It has truly a "wicked look," such as one sees in the alligators of children's picture books. The large head, bulging at the angles of the jaws, the glittering, yellow-irised eyes, and swiftly darting tongue constitute a truly forbidding front. (Grinnell and Grinnell, 1907)

Gerrhonotine lizards ("alligator lizards") have been a source of interest for over 300 years (e.g., Hernandez, 1651) and, as a group, show patterns of variation which make them ideal for the study of a number of aspects of lizard biology. They have been the subjects of investigations into biogeography, cytology, histology, breeding biology, habitat diversity, physiology, behavior, and feeding ecology, and have been shown to display interesting patterns in each.

Gerrhonotines comprise 38 currently recognized species distributed from southern British Columbia through the western United States, Mexico, and Central America, as far south as western Panama. Within the subfamily there is considerable diversity in body form and habitat, with species ranging from diminutive (SV=55 mm) to large (SV=over 200 mm) and from terrestrial desert forms to arboreal inhabitants of tropical cloud forests.

Unfortunately, phylogenetic relationships among gerrhonotine lizards remain uncertain and generic taxonomy is unstable despite numerous studies of the group (particularly Tihen, 1949a, b, 1954; Stebbins, 1958; and Waddick and Smith, 1974). In the following pages, I provide a discussion of variation in external characters in the subfamily, a review of previous analyses of gerrhonotine phylogeny, and a key to, and diagnoses of, all genera and species. I also reanalyze the variation in the subfamily with the purpose of postulating phylogeny. Finally, the taxonomic and biogeographic implications of this phylogeny are discussed.

The analysis of gerrhonotine relationships presented here is based on a detailed re-examination of external features, largely of scalation, in the subfamily. Emphasis on external features is necessitated by the existence of only alcoholic material for the majority of gerrhonotine species; several are known only from one or a few specimens, and no skeletal material. An analysis of the limited skeletal material available is presented elsewhere (Good, 1987b).

This analysis differs from previous works in that a more complete survey of taxa has been conducted, including examination of several forms unknown or unavailable to previous workers, and in that a cladistic approach to phylogenetic analysis has been taken (cf. Hennig, 1966). Rieppel (1980) and Gauthier (1982) used such an approach in analyses involving gerrhonotine taxa, but both focused on broader questions of anguimorph relationship and did not include all gerrhonotine species.

MONOPHYLY OF THE GERRHONOTINAE

In any analysis purporting to investigate phylogeny, it is essential that the taxa being analyzed form a monophyletic group; otherwise there is a danger of using as an outgroup something that is in fact more closely related to some of the taxa being studied than to others. Gerrhonotines are usually (e.g., Tihen, 1949a) described as anguid lizards with well developed limbs and a lateral fold. This is insufficient to diagnose a monophyletic group: limbs are certainly ancestral, and a lateral fold characterizes a larger group within the Anguidae, occurring also in the Anguinae and some Glyptosaurinae (McDowell and Bogert, 1954). Gauthier (1982) listed eight osteological characters as synapomorphies for the Gerrhonotinae: 1) relatively elongate supratemporal bones, 2) closely apposed palatines, 3) delicate, closely spaced teeth, 4) elongate anterior and posterior processes of the maxilla, 5) a closed premaxillary fenestra, 6) a weak dividing ridge on the dorsal surface of the supradental shelf of the maxilla, 7) a fused hourglass-shaped frontal, and 8) frontoparietal scales nearly in contact on the midline. However, character 5 is ancestral for the Anguidae (Rieppel, 1980) and character 8 is variable in all three extant anguid subfamilies. Characters 1 and 7, however, are derived features invariant in and restricted to the Gerrhonotinae among anguids, and characters 2, 3, 4, and 6, although variable in the subfamily, are probably also synapomorphies. The Gerrhonotinae is therefore appropriately considered to be monophyletic.

MATERIALS AND METHODS

Specimens were examined of all but one (*Elgaria parva*, for which excellent photographs and drawings were available) of the currently recognized gerrhonotine species (Appendix A) and detailed observations were made of their external morphology. Both inter- and intraspecific variation were noted and more-or-less discrete states were determined. There were varying degrees of correlation among characters but in those cases in which common causation was apparent only one of the correlated characters was used in the analysis.

Polarization of characters into ancestral and derived character states was accomplished through comparison with the anguid subfamilies Anguinae and Diploglossinae as outgroups. It is unclear which of the non-gerrhonotine subfamilies of anguids is most closely allied with the Gerrhonotinae; biochemical evidence weakly suggests the Anguinae (Good, 1987a), morphological evidence weakly suggests the Diploglossinae (Gauthier, 1982). Neither of the other two anguid subfamilies (sensu Gauthier, 1982), the Anniellinae nor the extinct Glyptosaurinae, was used in this analysis, the latter because of lack of sufficient comparative material and the former because, with its highly derived sand-swimming form, homologies with gerrhonotine scale conditions were difficult to establish.

If only one of the gerrhonotine character states was seen among the anguine or diploglossine species, it was assumed to be ancestral for the Gerrhonotinae. If more than one gerrhonotine state occurred elsewhere, determination of polarity in the Gerrhonotinae depended on the distribution of states among the outgroups: if alternate states were fixed in the Anguinae and Diploglossinae or if both were variable for more than one gerrhonotine state, polarity could not be determined and the character was included as unordered. If, however, either outgroup was fixed for a single gerrhonotine state while the other was polymorphic, the state shared by both outgroups was considered to be ancestral for the Gerrhonotinae (this was based on the assumption that the Diploglossinae and Anguinae are monophyletic taxa; see Gauthier, 1982). Polarity was impossible to determine if no gerrhonotine state occurred in either outgroup.

Subsequent to establishing polarities for as many characters as possible, hypotheses of relationships were determined first by inspection and then, in order to insure that alternative interpretations of character patterns were not missed, the analysis was re-done using

Swofford's (1985) PAUP computer program designed to compare large numbers of alternative branching diagrams in order to find the one that most parsimoniously explains character variation. Parsimony in this analysis refers to minimizing numbers of character state changes and assumes that the smaller the number of changes required by an hypothesis of relationships, the more likely that hypothesis is to be an accurate estimation of reality. PAUP was run with Farris optimization; global branch swapping; rooting at an imaginary, completely plesiomorphic ancestor; and characters unordered where polarity was uncertain.

In order to avoid much of what I saw as "noise" resulting from homoplasy in the analysis, I chose a 10% level for what I considered "significant"; any character state present in fewer than 10% of the specimens examined for a given species was disregarded in this analysis. The choice of a 10% significance level was a procedural one; acceptance of the presence of a character state in any percentage greater than zero completely swamped the analysis with homoplasies.

Certain characters showed two or more states in some species and were considered to be polymorphisms. Because of small sample sizes in many species, when the derived condition was seen in a polymorphic state it was scored simply as "present." The character states discussed in this paper are sometimes fixed and sometimes polymorphic, depending on the species; reference to the text should be made to determine the condition for each.

In some instances, published levels of occurrence of a character state differed from my observations. Personal observation was accepted over these published accounts.

Inconsistent head-scale nomenclature has led to confusion in the past. This analysis uses the system described by Waddick and Smith (1974), with certain modifications suggested by Campbell (1982). Any further modifications are discussed when they arise. Figure 1 illustrates this nomenclature on a schematic representation of the probable ancestral gerrhonotine morphology.

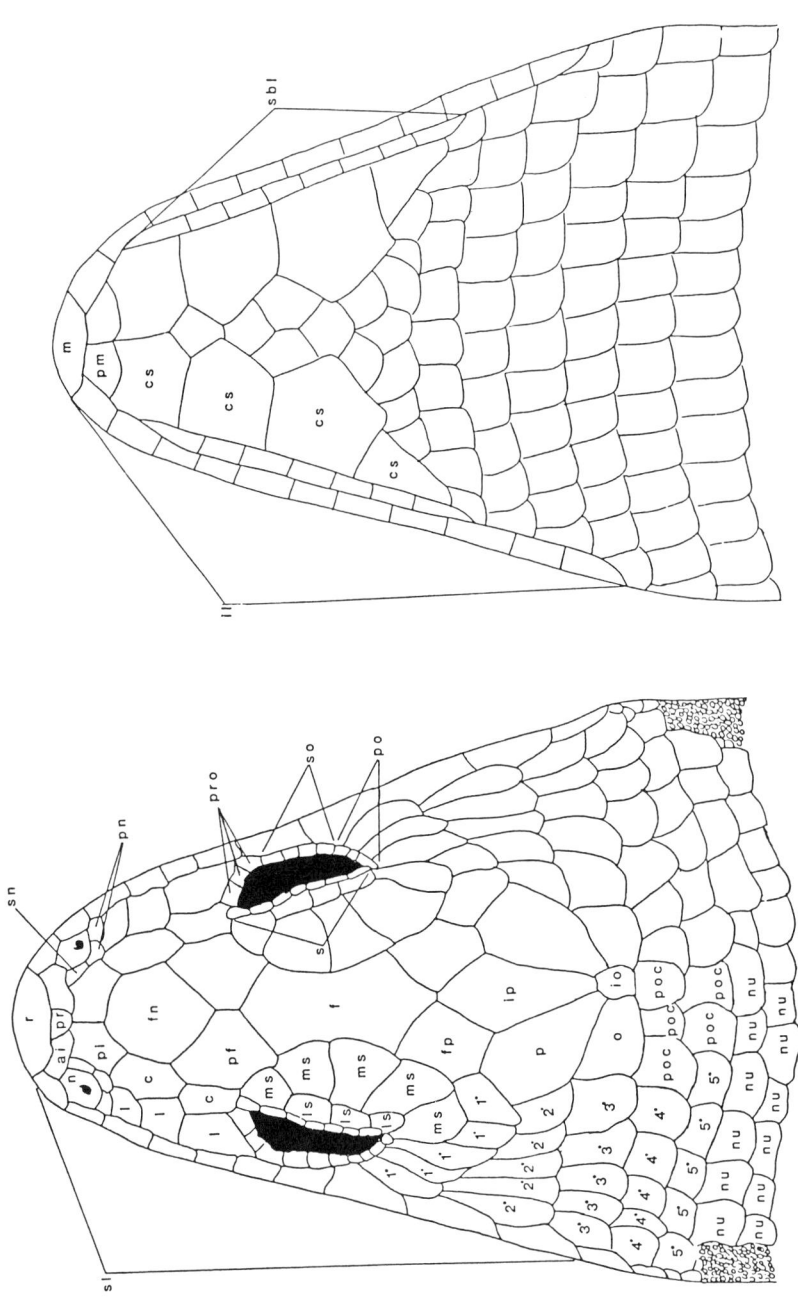

FIGURE 1. Head-scale nomenclature illustrated on the proposed gerrhonotine prototype (modified from Tihen, 1949a). ai=anterior internasal, c=canthal, cs=chinshield, f=frontal, fn=frontonasal, fp=frontoparietal, ip=interparietal, io=interoccipital, l=loreal, ls=lateral supraocular, m=mental, ms=medial supraocular, n=nasal, nu=nuchal, o=occipital, p=parietal, pf=prefrontal, pi=posterior internasal, pm=postmental, pn=postnasal, po=postorbital, poc=postoccipital, pr=postrostral, pro=preocular, r=rostral, s=superciliary, sbl=sublabial, sl=supralabial, sn=supranasal, so=suborbital, 1=primary temporal, 2=secondary temporal, 3=tertiary temporal, 4=fourth temporal, 5=fifth temporal.

CHARACTER ANALYSIS

The following is a description of variation in gerrhonotine external morphology. In each section the elements discussed are first defined and the ancestral condition postulated, based on outgroup comparison with the Diploglossinae and Anguinae. Variation in the character among gerrhonotine taxa is then discussed. The percentage of specimens examined showing a given state is provided in parentheses; where parentheses are absent, all specimens showed the trait.

The characters discussed here are summarized in Appendix B.

Rostral. The single medial scale anteriormost on the upper jaw. In the ancestral condition, contact is with the anterior supralabials, the anterior internasals, and the postrostral.

Nasal-rostral contact occurs in *Barisia rudicollis* and in *Elgaria* but homology is not suggested, since contact in *Elgaria* results from the loss of the anterior internasals (Figure 2b) while *B. rudicollis* retains these scales medial to the nasals (Figure 2c). Contact was reported also in a few *Gerrhonotus liocephalus* (1%), *Mesaspis monticola* (3%), and *M. moreleti* (1%) by Waddick and Smith (1974). They further reported a lack of contact in a few *E. coerulea* (2%) and *E. kingii* (1%), but they did not discuss the relationship between contact and internasal presence or absence in these species.

One of the two known specimens of *Abronia reidi* (the paratype) shows narrow contact of the rostral with the expanded supranasals; Waddick and Smith (1974) suggested that there may be contact between the rostral and the posterior internasals in some *A. taeniata* (6%; some or all of these might have been *A. graminea*, since these authors considered the two to be conspecific), a condition observed in the present study only in a few (<5%) *E. coerulea*. Waddick and Smith also stated that the anterior internasals sometimes fail to contact on the midline in *A. taeniata*. They were unclear as to whether this was due to posterior internasal-rostral contact or to the presence of a postrostral; it is here assumed that the former was the case since postrostrals are otherwise unknown in *Abronia*. None of these conditions, with the possible exception of the supranasal-rostral contact in *A. reidi*, occurs frequently enough to be considered significant using the criteria outlined in the previous chapter.

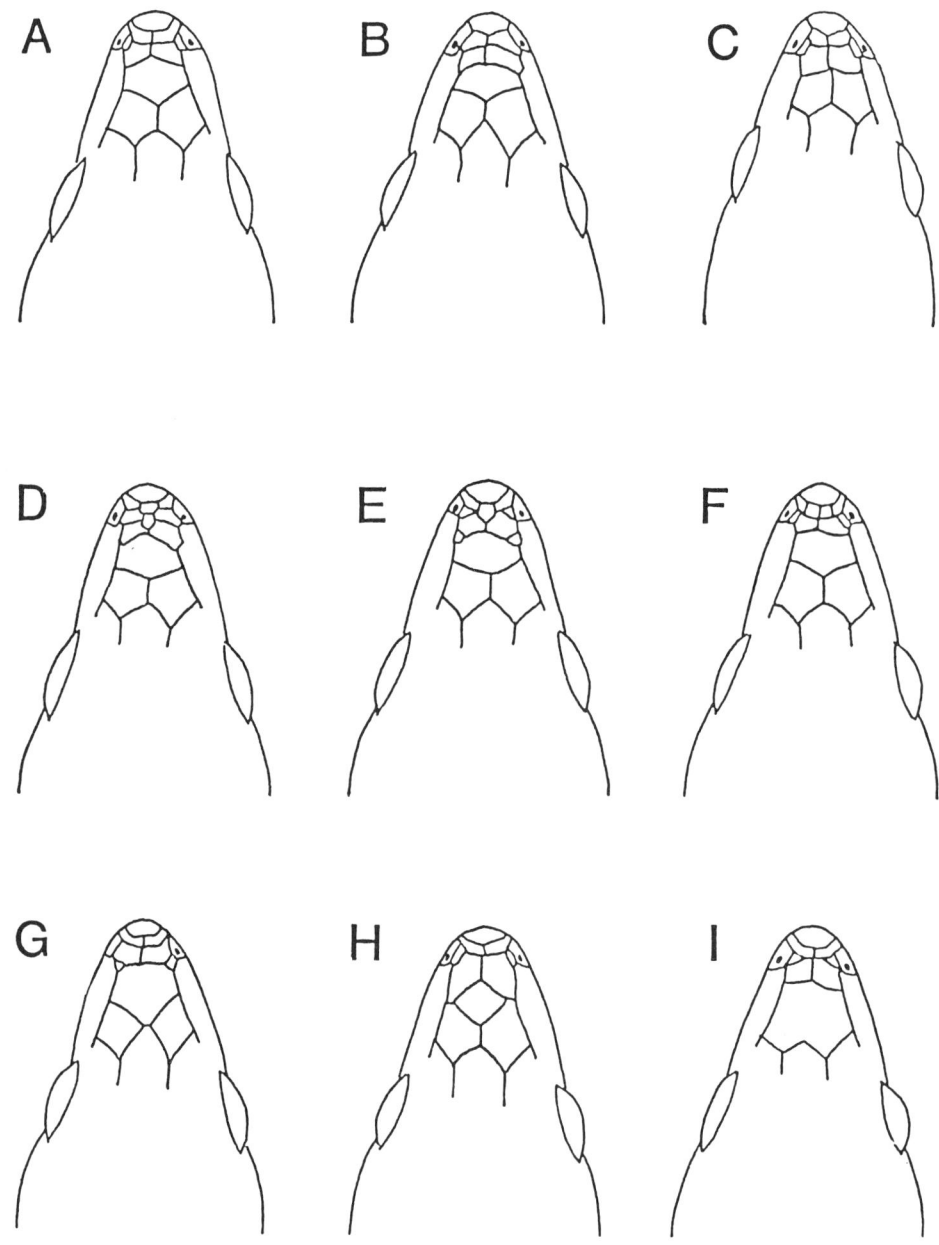

FIGURE 2. Schematic representation of dorsal snout scale variation in the Gerrhonotinae. A=general gerrhonotine condition, B=*Elgaria*, C=*Barisia rudicollis*, D=*Coloptychon rhombifer*, E=*Mesaspis antauges*, F=*Abronia bogerti*, G=*M. monticola*, H=*A. mixteca*, I=*M. moreleti*.

Postrostrals. The 1-2 longitudinally arranged medial snout scales posterior to the rostral. Contact is with the rostral and with various other elements of the snout, depending on the condition of those elements (see below). As suggested by Gauthier (1982), the presence of two postrostrals is a derived feature. However, because the outgroups vary in this character, whether a single postrostral or the absence of any such elements is ancestral for the Gerrhonotinae is unclear.

Tihen (1949a) illustrated the "gerrhonotine prototype" as lacking rostral-postrostral contact but gave no explanation for this belief. He also suggested without explanation that the postrostral in many forms probably arose through fusion of the middle two of the four anterior internasals, which he considered to be ancestral (although they probably are not; see below).

Two longitudinally arranged postrostrals are seen only in *Coloptychon rhombifer* (Figure 2d) in two of the three known specimens; the other lacks postrostrals altogether. Single postrostrals occur in *Gerrhonotus liocephalus*, *Mesaspis antauges*, and *M. juarezi* (44.5%) (Figure 2e). A few specimens (<10%) of *M. gadovii*, *M. viridiflava*, *M. monticola*, *M. moreleti*, *Elgaria panamintina*, and *E. coerulea* seen in this study also possessed postrostrals, and Waddick and Smith (1974) recorded low levels (1-7%) in most of the other gerrhonotines they examined except *Abronia*. They also reported 35.5% of their specimens of *M. gadovii* as having postrostrals, a percentage far higher than that seen in the present study. The reason for this discrepancy is unclear.

The loss of the postrostral in most gerrhonotines permits the midline contact of the anterior internasals.

Nasals. The paired anterolateral scales through which the external nares pass. Each is invariably a single scale that in the ancestral condition is in contact with the first two supralabials, an anterior internasal, a supranasal, and two postnasals.

Contact with the rostral is discussed above. In 9.5% of the *Mesaspis moreleti* specimens examined separation of the postnasals allowed for contact of the nasal and cantholoreal elements. Contact with the cantholoreal elements through fusion of various of these elements with the postnasals is discussed below.

In *Coloptychon rhombifer*, *Gerrhonotus liocephalus* (40%), *G. lugoi*, *Abronia graminea* (60%), *A. taeniata* (50%), *A. deppii*, *A. mixteca* (80%), and *A. oaxacae* (30%), the nasal extends posteriorly to contact the third supralabial (Figure 3b,e,g). It might be suggested that this contact is more likely in *Gerrhonotus* and *Coloptychon* than in *Abronia* to be the result of high supralabial number (see below), but supralabial reduction in the Gerrhonotinae occurs primarily at the posterior end of the series; there is little variation in the elements adjacent to the nasals.

The "prenasal" mentioned by Smith and Alvarez del Toro (1963) in *Abronia lythrochila* was seen in no gerrhonotine observed in the present study.

Anterior internasals. The paired scales immediately posterior to the rostral. In the ancestral condition the two anterior internasals are undivided and contact the rostral, the postrostral (if this element is ancestrally present), the supralabials, the nasals, the supranasals, and the posterior internasals. Tihen (1949a) suggested that four anterior internasals are ancestral but he gave no explanation for this hypothesis.

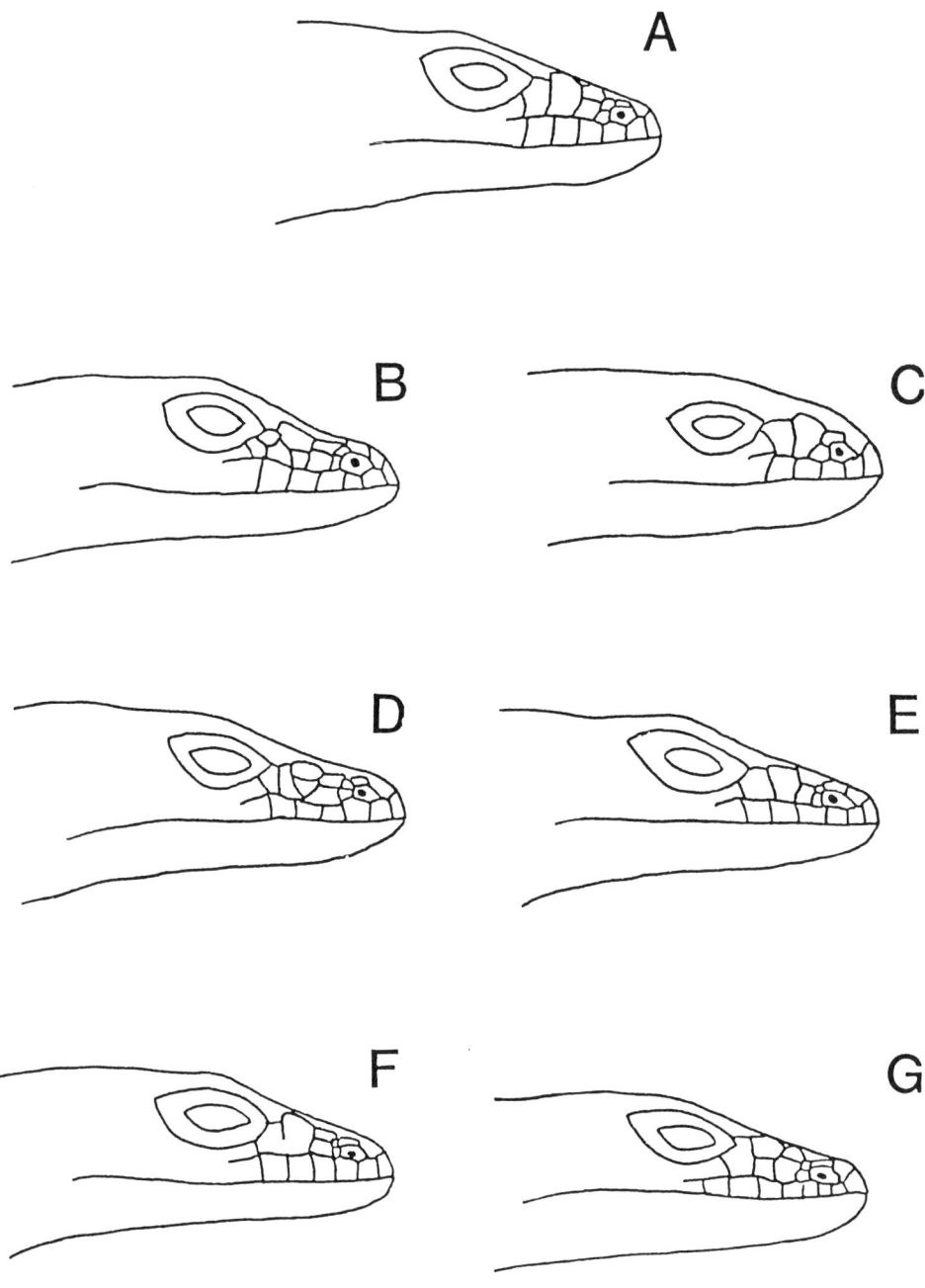

FIGURE 3. Schematic representation of lateral snout scale variation in the Gerrhonotinae. A=general gerrhonotine condition, B=*Coloptychon rhombifer*, C=*Barisia imbricata*, D=*Mesaspis moreleti*, E=*Abronia mixteca*, F=*M. antauges*, G=*Gerrhonotus lugoi*.

Anterior internasals are lacking from *Elgaria* (Figure 2b) and nasal-rostral contact precludes contact with the supralabials in *Barisia rudicollis* (Figure 2c). Anterior internasal absence in *Elgaria* is responsible for the nasal-rostral contact and supranasal expansion seen in the genus; the supranasals expand to replace the missing internasals. Waddick and Smith (1974) recorded rare (1%) anterior internasal presence in *E. coerulea*. They also recorded 9% of the *Mesaspis viridiflava* specimens they examined as not showing these scales.

Loss of the postrostral allows for midline contact of the internasals except in those individuals in which there is supranasal- or posterior internasal-rostral contact (see above). In *Abronia bogerti*, *Mesaspis viridiflava* (50%), and *M. juarezi* (22%), each anterior internasal is divided so that a transverse series of four scales is formed (Figure 2f). Other *M. juarezi* show this division on only one side (20%) as does one of the three known specimens of *Gerrhonotus lugoi*. Since the ancestral two internasals are seen in both other *G. lugoi* specimens, and since division is seen on only one side of this one, it is assumed here that two scales are characteristic of the species. Tihen (1954) suggested that the condition in the single known specimen of *A. bogerti* is probably due to individual variation rather than being a characteristic of the species. Although he stated that four anterior internasals are occasionally seen in other species of *Abronia*, no such instance was observed in the present study.

Posterior internasals. The second pair of large elements on the snout (except in those species in which the supranasals are expanded). They are unreduced, undivided scales in the ancestral condition, in contact with each other and with the postrostral (if one is ancestrally present), the anterior internasals, the supranasals, the upper postnasals, the cantholoreal series, and the frontonasal. The posterior internasals ancestrally approximate the anterior in size.

The usual interpretation of snout scale morphology in *Mesaspis monticola* has been that the supranasals are lost while the posterior internasals retain the ancestral condition (Tihen, 1949b). The presence of reduced scales medial to the postnasals (too far anterior to be part of the cantholoreal series) in 25% of the *M. monticola* specimens examined suggests that these scales are remnants of the posterior internasals; the elements usually considered posterior internasals are therefore best interpreted as expanded supranasals (Figure 2g). The remaining 75% of the *M. monticola* specimens lacked posterior internasals altogether; 18% of the *M. moreleti* specimens examined shared the reduction in these elements while in another 10% posterior internasals were absent. The posterior internasal is also missing from one side of the type of *Elgaria parva*, but since it is present on the other side of the specimen and on both sides of the other known specimen as well as in all other gerrhonotines except as noted above, this is probably not a characteristic of the species.

In *Abronia graminea*, *A. taeniata*, *A. deppii*, *A. mixteca*, and *A. oaxacae*, the posterior internasals have approximately doubled in size relative to the anterior internasals, not wholly as a result of the internasal-canthal fusion seen in those species (Figure 2h).

A small scale is present posterior to each posterior internasal in *Mesaspis antauges* (Figure 2e). This scale probably results from the division of the posterior internasal into a large anterior and a small posterior portion. The posterior internasal on one side of the holotype of *Abronia reidi* is also divided (Werler and Shannon, 1961), but since only two specimens are known, it is unclear whether this represents a polymorphism in the species or is an anomaly of the type specimen.

Fusion of the posterior internasals with the anterior canthal is discussed below.

Contact with the prefrontals occurs in those species in which the anterior canthal is missing (see below). Contact with the rostral occurs at varying frequencies in populations of *Elgaria coerulea*.

The "azygous internasal" in *Abronia chiszari* (Smith and Smith, 1981) was seen in no other gerrhonotine specimen. It may be the result of individual variation.

Supranasals. The paired elements immediately medial to the nasals. Gerrhonotines ancestrally possess small, unexpanded scales which contact the nasals, the upper postnasals, the posterior internasals, and the anterior internasals.

Medially expanded supranasals are seen in *Coloptychon rhombifer*, *Elgaria*, *Gerrhonotus liocephalus* (60%), *Abronia mitchelli*, *A. reidi*, *A. ornelasi*, *A. matudai* (one of four specimens examined), *Mesaspis antauges*, *M. juarezi*, *M. viridiflava*, *M. moreleti*, and *M. monticola* (Figure 2b,d,e,g,i). In *C. rhombifer* (two of three known specimens), *Elgaria*, *M. juarezi* (22.5%) *M. monticola*, *M. moreleti* (28%), *A. reidi*, and *A. ornelasi* this expansion allows for contact of the supranasals on the midline of the snout (Figure 2b,d,g). A few *E. kingii* (2%), *E. coerulea* (2%), and *E. multicarinata* (14%) were cited by Waddick and Smith (1974) as lacking contact, but this was not observed in the present study except in *E. coerulea* (<10%). Supranasal expansion is extremely variable in *M. moreleti*, which shows conditions ranging from a virtually complete lack of expansion to midline contact. Slight supranasal expansion is also seen on one side of the type of *A. chiszari*, but Smith and Smith (1981) considered this to be individual variation rather than a characteristic of the species.

Midline expansion of the supranasals in *Elgaria* probably results from the loss of the anterior internasals (see above). This loss is not seen in any other species, in which supranasal expansion usually occurs at the expense of the posterior internasals. This should probably not be considered homologous to the condition in *Elgaria*.

Posterolateral expansion of the supranasals through fusion with the upper postnasals is seen in *Barisia* (Figure 3c). In *Abronia*, only the holotype of *A. mixteca* shows fusion of the supranasal and both postnasals and then only on one side (Bogert and Porter, 1967).

The reduction of the posterior internasals discussed above in conjunction with the supranasal expansion seen in *Mesaspis monticola* and *M. moreleti* allows for contact of the supranasals with the frontonasal (Figure 2g). Contact is also seen in one of the two known specimens of *Abronia reidi* (the paratype) and in a few *Elgaria coerulea* (<10%). It is unclear whether this represents a polymorphism in *A. reidi* or an anomaly in the single specimen showing the derived condition. In either case, the contact in *A. reidi* should probably not be considered homologous to that in *Mesaspis*, since it does not result from a reduction in posterior internasals.

The supranasal was lacking from one side of one of the specimens of *Abronia ochoterenai* examined by Hartweg and Tihen (1946). However, as one of the specimens they examined may, in fact, have been *A. lythrochila* (Smith and Alvarez del Toro, 1963), it is not certain which species might occasionally lack this scale. In either case, since it occurs on only one side of the specimen, and since no other specimens of either species have been reported to lack supranasals, I will not consider it further. Hartweg and Tihen's observation that supranasals are also absent from *A. aurita* and Waddick and Smith's (1974) observation that they are lacking from a few *A. taeniata* (*A. graminea*?) are not substantiated by the present study. Hartweg and Tihen had never seen a specimen of *A. aurita* (Tihen, 1954), so their character descriptions for that species must be suspect. The type specimen of *A. montecristoi* lacks a supranasal on one side. Since no other gerrhonotine lacks supranasals, this is probably not a characteristic of the species. However, since only one specimen has been examined, this cannot be stated with certainty.

Postnasals. The pair of small scales contacting the posterior margin of the nasal. In the ancestral condition these scales contact each other as well as the nasal, the second and third supralabials, the supranasal, and the cantholoreal series.

Separation of the postnasals allowing the nasal to contact the cantholoreal series is discussed above, as is fusion of the upper postnasal with the supranasal. Fusion of the lower postnasal with the anterior loreal, again allowing for nasal-cantholoreal contact, is seen in *Barisia imbricata, B. levicollis, Mesaspis viridiflava, M. monticola,* and a few *M. moreleti* (4%) (Figure 3c).

In one of the three *Coloptychon rhombifer* specimens examined and in one of the three known *Gerrhonotus lugoi*, the lower postnasal is substantially larger than the upper. This condition is not seen in any other gerrhonotine.

Contact with the fourth supralabial is sometimes seen in those species in which the nasal reaches the third (see above).

Frontonasal. The single medial scale posterior to the posterior internasals and medial to the cantholoreal series. Contact with the posterior internasals, the cantholoreal series, the frontal, and the prefrontals is ancestral. Frontonasal characteristics were among the few scale characters emphasized by Meszoely (1970) in his review of fossil anguids.

Lack of a frontonasal is seen in *Barisia imbricata* (97%), *B. levicollis, B. rudicollis, Mesaspis antauges* (one of three specimens examined), *M. juarezi* (40%), *M. viridiflava,* and *Abronia oaxacae* (33%) (Figure 2c). Most other *A. oaxacae* specimens examined had reduced frontonasals. *Mesaspis gadovii* also sometimes (8%) lacks a frontonasal. Martin del Campo (1939), Hartweg and Tihen (1946), and Tihen (1954) observed the loss of a frontonasal in some *A. ochoterenai*, but there is some question about the identification of the specimens seen by the latter authors as *A. ochoterenai* or *A. lythrochila* (Smith and Alvarez del Toro, 1963). Boulenger (1885) and Hidalgo (1983) also cited absence as occasionally occurring in *A. aurita*.

In those species retaining the frontonasal, contact with the frontal is seen in *Mesaspis gadovii* (47%), *M. juarezi* (30%), *M. moreleti* (96%), *M. monticola* (67%), *Abronia reidi* (one of the two known specimens), *A. ornelasi, A. aurita* (one of four specimens examined), *A. taeniata* (25%), and *A. deppii* (60%). Hartweg and Tihen (1946) observed contact in two of the four specimens of "*A. ochoterenai*" they examined (but see Smith and Alvarez del Toro, 1963). Bogert and Porter (1967) described the holotype of *A. mixteca* as exhibiting contact but none of the specimens examined in this study showed it. Contact is also rarely seen in *Elgaria coerulea* (Tanner, 1959) as is division of the frontonasal into two scales (Van Denburgh, 1922).

Contact of the frontonasal with the supranasals is discussed above. Contact with the anterior canthal is lost when that scale disappears and contact with the medial supraoculars occurs when the prefrontals are lost (see below).

Cantholoreal series. The series of 1-5 elements comprising the lateral aspect of the snout posterior to the postnasals and anterior to the preoculars. In the ancestral condition they contact the supralabials, the postnasals, the posterior internasals, the frontonasal, the prefrontals, the medial supraoculars, the anterior superciliaries, and the preoculars. Determination of the ancestral state among gerrhonotine cantholoreal patterns is difficult as all anguines and many diploglossines have more scales in this region than does any gerrhonotine. However, this suggests that "many" may be ancestral.

The interpretation of homologies among cantholoreal elements is often difficult and in some groups highly questionable. This is especially true in such species as *Mesaspis moreleti*, which shows virtually the entire range of variation seen in the Gerrhonotinae. However, examination of a large number of specimens illustrating a continuum of forms yields useful hypotheses.

The gerrhonotines showing the highest number of cantholoreal elements are *Coloptychon rhombifer* and *Gerrhonotus liocephalus* (70%) which possess a row of two canthal elements and a row of three loreals (Figure 3b). In one of the three known specimens of *C. rhombifer*, the posterior canthal is fused with the middle loreal on both sides. It is unclear with the specimens available whether this is due to a polymorphism or to an anomaly in this single specimen. Loss of one of the loreal elements was seen in another 11% of the *G. liocephalus* specimens and a similar pattern is seen in *G. lugoi*, *Mesaspis monticola* (22%), *M. moreleti* (14.5%), and *M. juarezi* (<10%; Karges and Wright, 1987) (Figure 3d,g). However, since undivided cantholoreals predominate in these species, the posterior elements in the three *Mesaspis* are almost certainly derived from a divided single cantholoreal. Tihen (1954) also observed this condition in a single specimen of *M. gadovii*, but no such specimens were observed in this study. The type of *G. liocephalus taylori* (Tihen, 1954) shows a fusion of the anterior canthal with the posterior internasal on one side.

The basic theme in all other gerrhonotines is to have an anterior canthal, an anterior loreal, and a single posterior cantholoreal (Figure 3a). Variations on this are: 1) Fusion of the cantholoreal with the anterior canthal is seen in *Mesaspis viridiflava*, *Barisia imbricata*, *B. levicollis*, and in one of the two known specimens of *B. rudicollis* (the holotype) (Figure 3c). The other *B. rudicollis* specimen shows a fusion of the anterior loreal with the cantholoreal and has a free canthal. 2) Fusion of the anterior loreal with the lower postnasal is described above. 3) Loss of the anterior canthals through fusion with the posterior internasals is seen in *Abronia reidi*, *A. ornelasi* (one side of one of three specimens), *A. lythrochila* (75%), *A. taeniata* (90%), *A. graminea* (85%), *A. deppii*, *A. mixteca*, and *A. oaxacae* (Figure 3e). Anterior canthals in *A. taeniata* and *A. graminea* usually occur on only one side of the head and their presence is not considered characteristic. The anterior canthal is also lacking on one side of the type of *A. matudai* and Campbell (1982) stated that *A. mixteca* sometimes has a canthal. Tihen (1954) suggested that canthals might also be lacking from *A. aurita* but he never examined a specimen. Hidalgo (1983) also cited rare absence in this species (he examined seven). No specimens observed here showed such a lack and it is unlikely that it is characteristic of the species. This is doubtless also true of the observations of rare canthal loss by Tihen (1954) in *A. ochoterenai* and Lynch and Smith (1965) in *Mesaspis antauges*. 4) The extension of the anterior loreals posteriorly to contact the preocular, excluding contact of the posterior cantholoreal and the supralabials occurs in *M. moreleti* (68%), *M. monticola* (66%), and a few *M. gadovii* (8%) (Figure 3d). 5) Some tendency to partial fusion of the posterior cantholoreal with the preocular is seen in *M. antauges* (Figure 3f). 6) A particularly high degree of variation is seen in cantholoreal morphology in *Elgaria coerulea*, *M. juarezi*, *M. viridiflava*, *M. moreleti*, *M. monticola*, and perhaps *M. antauges*, although too few specimens exist of the latter form to make such a statement certain.

According to Waddick and Smith (1974), *Abronia oaxacae* and some *A. taeniata* (*A. graminea*?) lack a cantholoreal. It is unclear whether they mean that the scale has been lost completely or divided into canthal and loreal elements. Neither condition was observed in this study.

Prefrontals. Paired dorsal elements between and lateral to the frontonasal and the frontal. In the ancestral state, contact is with the frontonasal, the cantholoreal series, the medial supraoculars, and the frontal. Contact with each other on the midline is derived.

Separation of the prefrontals from each other through contact of the frontal and frontonasal is described above. Contact with the posterior internasals occurs when the anterior canthal has been lost (see above). Prefrontal-superciliary contact is seen in *Coloptychon rhombifer*, *Gerrhonotus liocephalus* (50%), *Abronia reidi*, and on one side of the type of *A. chiszari* (Figure 4b). This condition was also observed by Tihen in a few *Mesaspis moreleti* (1949b) and *M. gadovii* (1954).

In 27% of the *Mesaspis moreleti* specimens examined, prefrontals were lacking through fusion with the frontonasal (Figure 2i). This was also seen in some *M. monticola* by Tihen (1949b, 1954) but was not observed in the present study. One specimen of *M. moreleti* cited by Dunn and Emlen (1932) was described as having a single prefrontal (=prefrontals fused?). This was also not observed here. Karges and Wright (1987) described occasional fusion of the prefrontals on the midline in *M. juarezi*.

Frontal. The single dorsal element posterior to the prefrontals and separating the medial supraoculars on either side of the head. In the primitive state the frontal contacts the prefrontals, the frontonasal, the medial supraoculars, the frontoparietals, and, narrowly, the interparietal.

Contact with the frontonasal is described above. Contact with the interparietal is lost in *Abronia kalaina* through midline contact of the frontoparietals (Figure 5b). Narrow contact of the frontoparietals is also seen in a single specimen of *Gerrhonotus lugoi*, but the small sample available makes it impossible to say whether this reflects a polymorphism in the latter species. *Abronia kalaina* also shows almost complete fusion of the frontal with the frontoparietals (Figure 5b).

The primitively narrow contact of the frontal and the interparietal is broadened in *Mesaspis antauges* and *M. juarezi* (Figure 5c). The "broad" contact described by Hartweg and Tihen (1946) for *M. moreleti* and by Smith and Smith (1981) for *Abronia chiszari* does not approach the condition in these species. In one of the three known specimens of *Coloptychon rhombifer* the frontal is transversely divided into two scales.

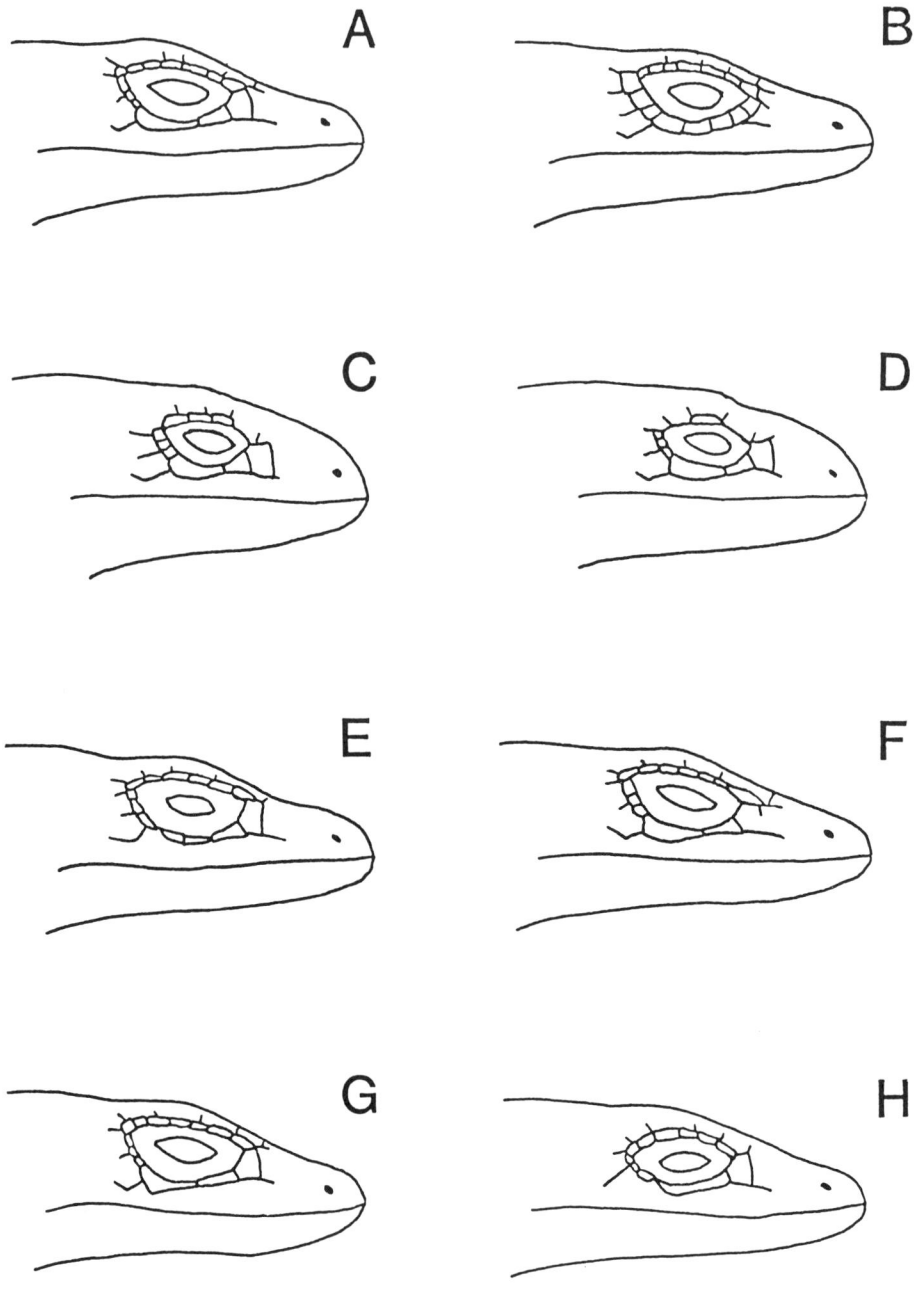

FIGURE 4. Schematic representation of circumocular scale variation in the Gerrhonotinae. A=general gerrhonotine condition, B=*Coloptychon rhombifer*, C=*Barisia imbricata*, D=*B. levicollis*, E=*Abronia salvadorensis*, F=*Mesaspis antauges*, G=*Elgaria kingii*, H=*A. deppii*.

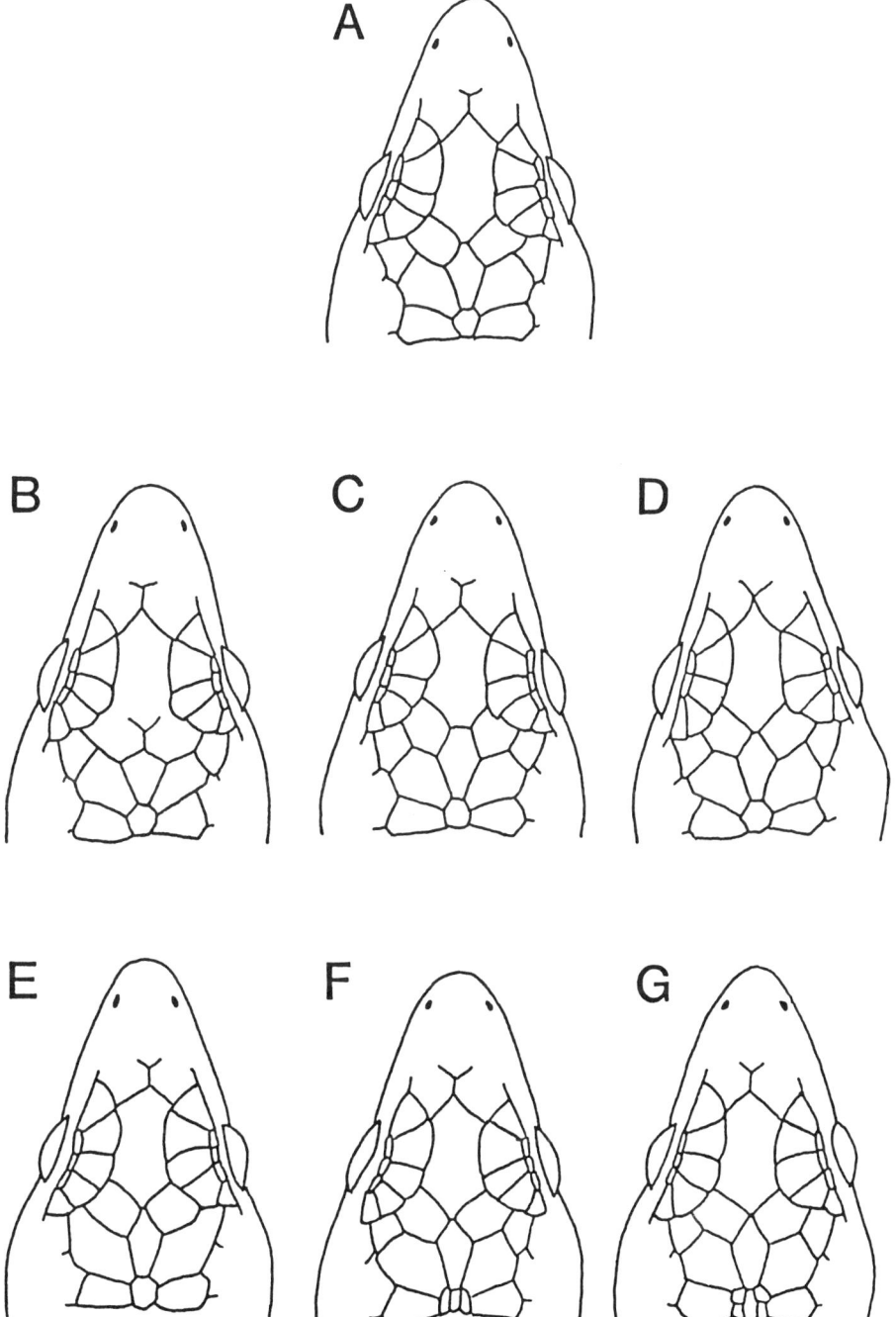

FIGURE 5. Schematic representation of dorsal head scale variation in the Gerrhonotinae. A=general gerrhonotine condition, B=*Abronia kalaina*, C=*Mesaspis antauges*, D=*M. monticola*, E=*A. lythrochila*, F=*A. mixteca*, G=*A. montecristoi*.

The frontal shows a sunken appearance in *Barisia imbricata* and *B. levicollis*, with the surrounding scales of the head protruding around it to give a concave appearance to this region of the head.

Medial supraoculars. The five scales in the medial of the two rows covering the dorsal aspect of the orbit. The medial supraoculars ancestrally contact the cantholoreal series, the superciliaries, the lateral supraoculars, the prefrontals, the frontal, the frontoparietals, the uppermost anterior temporals, and the postoculars.

Contact with the cantholoreal series is lost when the superciliaries contact the prefrontal (see above). In species in which the upper anterior temporals are reduced or lacking, the parietals come into contact with the medial supraoculars (see below). Frontonasal-supraocular contact occurs when the prefrontals are missing (see above).

Lateral supraoculars. The scales in the more lateral of the two rows covering the dorsal aspect of the orbit. The lateral supraoculars ancestrally consist of three scales in contact with the medial supraoculars and the superciliaries.

The ancestral condition of three lateral supraoculars was seen in all specimens except one of the three known *Coloptychon rhombifer* (four scales on one side and five on the other), *Elgaria coerulea* (30% with two scales), *Abronia matudai* (two and four scales each on one side of single specimens), *A. aurita* (four scales on one side of two of the four specimens examined), *Mesaspis juarezi* (20%), *M. viridiflava, M. monticola, M. moreleti* (36%), and *A. bogerti* (one side of the only known specimen), each with two scales. Campbell (1982) referred to two of six specimens of *A. ornelasi* as possessing two scales but this was not observed here. Gehlbach and Collette (1957) cited four scales in *A. oaxacae* and Tihen (1949b, 1954) cited specimens of *M. monticola* with three. The four scales cited by Stejneger (1907) for *M. monticola* are probably the result of a difference in interpretation of lateral supraoculars and superciliaries. Guillette and Smith (1982) observed four supraoculars in a single *Barisia imbricata*.

In the case of all of the above species except *Elgaria coerulea, Mesaspis juarezi, M. viridiflava, M. moreleti,* and *M. monticola* (Figure 5d), two or four scales were usually seen on only one side of any given specimen; the other side usually retained the primitive three scales. Three scales are therefore here considered characteristic of these species. One of three known *Coloptychon rhombifer* specimens showed 4-5 scales on both sides. Whether this is characteristic of the species is not discernable from the small sample available.

Superciliaries. The row of scales immediately above the eye and lateral to the supraoculars. They ancestrally consist of a complete row of 4-7 subequal scales in contact with the cantholoreal series, the upper preocular, the medial and lateral supraoculars, and the postoculars.

The superciliaries are reduced in *Barisia imbricata* and *B. rudicollis* usually to three scales extending from the primitive cantholoreal contact posterior to approximately the middle of the eye (Figure 4c). They are further reduced in *B. levicollis* to a single scale at the midpoint of the orbit (Figure 4d). Complete loss of superciliaries has been observed in a single individual of *B. imbricata* (Tihen, 1949b; Guillette and Smith, 1982). Cantholoreal contact is lost in *Abronia salvadorensis, A. deppii* (43%), *A. oaxacae,* and a few *A. graminea* (4%), but without the general reduction seen in *Barisia* (Figure 4e,h).

The anteriormost superciliary is elongate in *Mesaspis antauges, M. juarezi, Abronia chiszari,* and *A. bogerti* (Figure 4f). Tihen (1954) stated that the anterior superciliary is also elongate in *A. oaxacae,* but this does not seem to be the case. Contact with the prefrontals is discussed above.

Reduction of the posterior superciliaries was cited by Gauthier (1982) as diagnostic of *Abronia,* but this is mistaken.

Preoculars. The scales immediately anterior to the orbit. They ancestrally contact the supralabials, the cantholoreal series, the suboculars, and the superciliaries. The ancestral number of preoculars is unclear, as anguines have several scales undifferentiated from the suboculars while diploglossines have a single, well differentiated element.

The only gerrhonotines with multiple undifferentiated preoculars are *Coloptychon rhombifer, Gerrhonotus lugoi,* and *G. liocephalus* (87%) (Figure 4b). All others show a single element. The partial fusion of this single preocular with the cantholoreal in *Mesaspis antauges* is described above. In 25.5% of the *M. moreleti* specimens examined, the preocular appeared to be transversely divided into anterior and posterior halves (Figure 3d) but the complexity of cantholoreal variation in this species makes this interpretation suspect.

Suboculars. The scales making up the ventral edge of the orbit. The homology of anguine scales with gerrhonotine scales is unclear, but if the anguine lorilabials correspond to sublabials, it is likely that possession of a high number is the ancestral state; diploglossines vary in subocular number. As discussed above, determination of the polarity of preocular-subocular differentiation is impossible; the same is true of postocular-subocular differentiation. Contact is ancestrally with the supralabials, the preoculars, the postoculars and the lower primary temporals. Gauthier's (1982) view that there is an ontogenetic change in subocular-temporal contact in *Abronia* is mistaken; no such change was observed in any species for which both juvenile and adult specimens were available.

Coloptychon rhombifer shows the highest number of scales (3-5 in the region occupied by suboculars in species with a clear subocular-preocular and subocular-postocular distinction; Figure 4b); *Gerrhonotus* and most species with well differentiated suboculars have two (Figure 4a,c,d,f,g). Exceptions to this are: 1) *G. lugoi, Elgaria coerulea* (20%), *Abronia ornelasi* (one of three specimens examined), *A. ochoterenai* (one of two specimens examined), *A. matudai, A. lythrochila, A. vasconcelosii* (both sides of one and one side of the other specimen examined), *A. aurita* (both sides of three and one side of the fourth specimen examined), *A. salvadorensis, A. montecristoi* (one side of the only specimen examined), and a few *A. graminea* (6%) all have three scales (Figure 4e). Two scales were seen in one of the two *A. lythrochila* specimens examined by Smith and Alvarez del Toro (1963) and Werler and Shannon (1961) recorded three on one side of one of the only two known specimens of *A. reidi* (the holotype). Tihen (1949b) stated the same about the holotype of *Barisia rudicollis.* Other records of the rare occurrence (<5%) of three suboculars are in *G. liocephalus* (Tihen, 1954), *E. panamintina* (Stebbins, 1958), *Mesaspis monticola* (Tihen, 1949b), *M. juarezi* (Karges and Wright, 1987) and *B. levicollis* (Guillette and Smith, 1982). Of the above species, this analysis suggests that three suboculars are characteristic of *G. lugoi, E. coerulea, A. ochoterenai, A. matudai, A. lythrochila, A. vasconcelosii, A. aurita, A. salvadorensis,* and probably *A. montecristoi.*

However, this is based partly on corroboration by other characters; as this character itself stands, the conditions in *A. ornelasi*, *A. ochoterenai*, and *A. montecristoi* are particularly unclear. 2) A single subocular is present in *A. deppii* and *M. gadovii* and on one side of the types of *A. mitchelli* and *A. chiszari* (Figure 4h). Sanchez-Herrera and Lopez-Forment (1980) counted two suboculars in some *A. deppii* and Tihen (1954) observed the same in *M. gadovii*. Neither showed this condition in the present study. Tihen (1948) also counted one subocular in a specimen of *G. liocephalus*.

The lack of subocular-temporal contact usually cited for *Gerrhonotus* is the result of the presence of unreduced postoculars, the lowest of which causes separation; this scale corresponds to the posterior subocular in *Coloptychon rhombifer*. Among the species with reduced postoculars, subocular-temporal contact has been lost in *Elgaria parva* and in all *Abronia* except *A. mitchelli*, *A. reidi*, and *A. ornelasi* (Figure 4e,h). Some contact was observed in *A. graminea* (45%, a percentage much higher than the 3% observed by Tihen, 1954, in "*A. taeniata*") as well as on one side of the only known specimen of *A. kalaina*.

There is a weak tendency in *Elgaria kingii*, *E. panamintina*, *E. cedrosensis*, and *E. paucicarinata* for the posterior subocular to be rather triangular in shape (Figure 4g).

Postoculars. The vertical series of 2-6 scales at the posterior margin of the orbit, isolating it from contact with the anterior temporals. Contact is with the suboculars, the anterior temporals, the superciliaries and the medial supraoculars. No significant variation was observed in this series except in the degree of subocular-postocular differentiation and in the loss of superciliary contact discussed above.

Temporal series. The series of scales on the side of the head between the eye and the ear. Anguines and diploglossines possess more vertical rows of temporals than does any gerrhonotine; a high number is therefore probably ancestral. The ancestral number of scales per row is 5-6. They are bordered above by the frontoparietals, the parietals, and the occipitals. Below they contact the supralabials. All temporal scales are ancestrally more-or-less equal in size.

The only gerrhonotines showing the ancestral condition of 5-6 scales in all temporal rows are *Coloptychon rhombifer* and one specimen of *Gerrhonotus lugoi* (the other two are similar to *G. liocephalus*). All other species have four scales per row except as described below:

The fifth temporal row (that nearest the ear) is lost in all *Abronia* (Figure 6b,c,d,e,f). The fourth row has been reduced in all *Abronia* except *A. mitchelli*, *A. reidi*, and *A. ornelasi* (Figure 6c,e), and completely lost in *A. lythrochila*, *A. vasconcelosii*, *A. aurita*, *A. oaxacae*, and some *A. mixteca* (20%) (Figure 6d,f).

Five scales are present in the third temporal row in many *Barisia imbricata* (73%) and *B. levicollis* (60%). Smith and Alvarez del Toro (1963) described *Abronia ochoterenai* as having 5-6 tertiary temporals, but this seems to be due to confusion as to the identity of certain scales. The third row has been reduced to three scales in *A. chiszari* and *A. bogerti* and to two scales in *A. oaxacae* (Figure 6e). However, the reduction in the former two species is due to an expansion of the first two temporal rows at the expense of the third (see below), while in the latter it stems from an increased size in all temporal scales, including those of the third row. The third row is also disrupted by expansion of the second in *A. aurita*, but this species retains the ancestral four scales (Figure 6f).

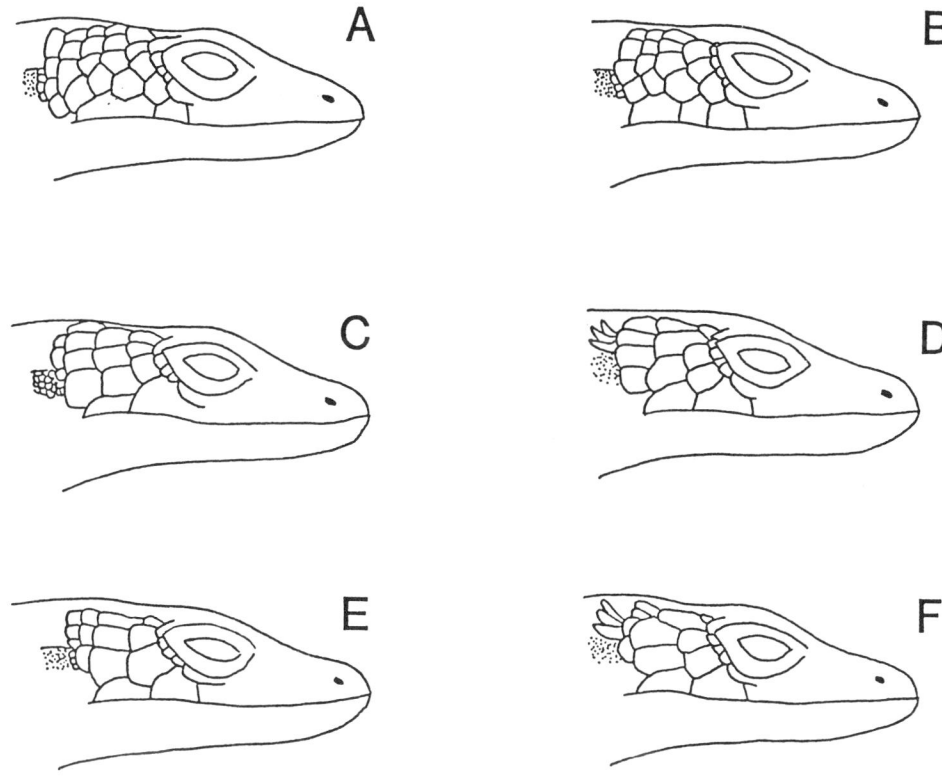

FIGURE 6. Schematic representation of temporal scale variation in the Gerrhonotinae. A=general gerrhonotine condition, B=*Abronia reidi*, C=*A. deppii*, D=*A. lythrochila*, E=*A. chiszari*, F=*A. aurita*.

The second row of temporals has been reduced to three scales in *Abronia ochoterenai*, *A. lythrochila*, *A. vasconcelosii* (both sides of one and one side of the other specimen examined), *A. aurita*, *A. salvadorensis*, *A. montecristoi* (one side of the only known specimen), *A. chiszari*, *A. bogerti*, *A. kalaina*, *A. fuscolabialis*, *A. graminea*, *A. taeniata* (92%), *A. deppii*, *A. mixteca* (40%), and *A. oaxacae* (67%) (Figure 6c,d,e,f). The remaining 33% of *A. oaxacae* specimens possessed only two scales in this row. Five secondary temporals are occasionally present in *Mesaspis juarezi* (Karges and Wright, 1987).

Modifications of the first temporal row are as follows: 1) In one of the two *Abronia ochoterenai* specimens examined, one of the upper scales had divided so that five anterior temporals were present on each side. 2) Three scales are present in *A. reidi* (four on one side of one of the two known specimens: the type), *A. matudai* (one side each of two of the four specimens examined, with the other side of one of these being reduced to two), *A. lythrochila* (both sides of two and one side of another out of four specimens examined), *A. vasconcelosii* (one of the two specimens examined, with the other one having two scales),

A. aurita (both sides of one and one side of another of the four specimens examined), *A. chiszari*, *A. deppii* (86.5%, with another 6.5% reduced to two), *A. mixteca* (20%), *A. oaxacae*, and rarely (<5%) in *A. graminea* (Figure 6b,c,d,e,f). Tihen (1954) stated that *A. ochoterenai* also has 3 primary temporals, but this was not observed in the present study. However, Smith and Alvarez del Toro (1963) also cited three scales in some specimens (range 3-5). An insignificant level (<10%) of modification to three or five scales has also been seen in *G. liocephalus* (Tihen, 1948), *E. kingii* (Webb, 1962), *Barisia imbricata* (Anderson and Lidicker, 1963), and *Mesaspis juarezi* (Karges and Wright, 1987). *Abronia bogerti* shows a further reduction to two primary temporals.

This reduction to 2-3 primary temporals has been accomplished in a variety of ways: in *Abronia reidi*, *A. matudai*, *A. lythrochila*, *A. vasconcelosii*, *A. aurita*, *A. oaxacae*, and *A. mixteca*, it is accomplished through reduction in the upper elements (Figure 6b,d,f) while in *A. deppii*, the lowest is fused with the penultimate supralabial (Figure 6c). *Abronia chiszari* and *A. bogerti*, in addition to showing reduction in the upper temporals, have the lower elements expanded to the exclusion of certain of the other elements (Figure 6e). This lower element expansion is also responsible for the reduction in tertiary temporal number discussed above, as well as the decreased number of primary elements contacting the orbit and the contact of the penultimate supralabial with the orbit, both discussed below.

In the case of upper element reduction, the uppermost (fourth) element is lost in *Abronia reidi* and *A. matudai* (Figure 6b); the third element is lost in *A. aurita*, *A. vasconcelosii*, *A. lythrochila*, *A. oaxacae*, and *A. mixteca* (Figure 6d,f). Often, when this latter state occurs, the secondary temporals are able to contact the medial supraoculars between the second and fourth primaries. Reduction of upper anterior temporal elements allows for the contact of the parietals with the medial supraoculars (thus excluding the temporals from the frontoparietals) in *A. reidi*, *A. matudai* (three out of four specimens examined), *A. lythrochila*, *A. vasconcelosii*, *A. aurita*, *A. chiszari*, and *A. bogerti* (Figure 5e). Hartweg and Tihen (1946) observed rare contact in *A. ochoterenai* but there is some confusion about the identity of one of their specimens (Smith and Alvarez del Toro, 1963). Smith and Alvarez del Toro (1962) recorded two specimens of *A. lythrochila* (identified by them as *A. ochoterenai*) as lacking contact. Although all specimens of *A. lythrochila* examined in this study showed this contact, it was seen on only one side of one of the two specimens seen by Smith and Alvarez del Toro. Hidalgo (1983) stated that contact is also sometimes lacking in *A. aurita* and *A. vasconcelosii*. However, one of the "*A. vasconcelosii*" specimens he examined (LACM 75514) was in fact *A. matudai*. Reduction or expansion of the lower anterior temporal elements limits the number of scales contacting the orbit to one in *A. bogerti*, *A. chiszari*, and *A. deppii*. *Coloptychon rhombifer*, *Gerrhonotus lugoi*, *G. liocephalus* (20%), *A. reidi*, and *A. ornelasi* show the ancestral condition of having three primary temporals in contact with the orbit.

The contacts of certain primary with certain secondary temporal elements have been used to differentiate species of gerrhonotines. These are largely the result of the character states described above and hence not independent characters.

Frontoparietals. The paired scales between and lateral to the frontal and the interparietal. Ancestral contact is with the frontal, the interparietal, the medial supraoculars, the parietals, and the upper anterior temporals.

Variation in the contacts of the frontoparietals with each other and with the temporals, and fusion with the frontal are discussed above.

Interparietal. A single medial element posterior to the frontal and anterior to the interoccipital. Contact is with the frontal, the frontoparietals, the parietals, and the interoccipital.

The degree of contact with the frontal is discussed above.

Parietals. Paired elements flanking the interparietal. Contact is ancestrally with the frontoparietals, the interparietal, the occipitals, the interoccipital, and the upper scales of the first and second temporal rows.

Aside from the occasional contact with the medial supraoculars due to the temporal reduction discussed above, little variation has been observed. Contact with the upper tertiary temporal was seen in one specimen of *Coloptychon rhombifer* (33%), but this may be an individual variant.

Interoccipital. The medial element posterior to the interparietal. It is ancestrally a single element in contact with the interparietal, the parietals, the occipitals, and the postoccipitals.

The interoccipital is divided into two scales in *Abronia mitchelli*, into three in *A. oaxacae* and *A. mixteca* (Figure 5f), and into five in *A. montecristoi* (Figure 5g). Three interoccipitals are also sometimes seen in *Gerrhonotus lugoi* and *Elgaria coerulea* (30%).

The usual number of interoccipitals in *Gerrhonotus lugoi* is open to question because all of the known specimens differ in the character. With only three specimens available, it is impossible to know whether this high variability is characteristic of the species.

Occipitals. The paired elements flanking the interoccipital. They contact the parietals, the interoccipital, the temporal series, and the postoccipitals. No significant variation has been observed in these elements.

Postoccipitals. The scales in transverse rows between the interoccipital and the nuchals. Postoccipitals ancestrally occur in two rows and contact the interoccipital, the occipitals, the temporals, and the nuchals. The scales are ancestrally smooth.

The number of postoccipital rows has been reduced to one in *Abronia mitchelli* and increased to three in *A. deppii* and *A. mixteca*, though Bogert and Porter (1967) suggested that only *A. mixteca* has three rows. Bogert and Porter also reported that *A. oaxacae* has a single row of postoccipitals but, as Campbell (1982) pointed out, this is not characteristic of the species. Campbell (1984) observed only one row in one of the seven *A. ornelasi* specimens he examined, but this was not observed here. The ancestral relative smoothness of these scales has been replaced in adult *Barisia* and *A. graminea, A. taeniata, A. deppii, A. mixteca,* and *A. oaxacae* by rugose scales (Figure 7). The rugosity mentioned by Smith and Alvarez del Toro (1963) in *A. lythrochila* does not compare with that in these species. In *A. kalaina, A. fuscolabialis, A. graminea,* and *A. taeniata,* and to a greater extent in *A. deppii, A. mixteca,* and *A. oaxacae* (Figure 7), the posterior scales of the head have developed characteristically knoblike scales. Strong keeling is present on the posterior part of the head of *B. rudicollis*.

Supra-auriculars. The small scales immediately dorsal to each ear. These scales are ancestrally smooth and undifferentiated.

The supra-auriculars have become characteristically protuberant in *Abronia ochoterenai, A. matudai, A. lythrochila, A. vasconcelosii,* and *A. aurita* (Figure 6d,f). There is an

FIGURE 7. Photograph of a 106 mm SV preserved specimen of *Abronia oaxacae* (MVZ 144197).

ontogenetic component to the degree of protuberance of these scales, with juveniles showing only slight expansion.

Pre-auriculars. The area of small scales anterior to the ear and posterior to the temporals and supralabials. These scales are ancestrally nongranular and thus considerably larger than the granulars of the side of the neck.

A reduction has occurred in the size of the pre-auriculars in *Abronia ochoterenai*, *A. matudai*, *A. lythrochila*, *A. vasconcelosii*, and *A. aurita* (Figure 6d,f). This tendency is also seen in *A. taeniata* (80%) and *A. graminea* (85%).

Supralabials. The 8-14 scales making up the dorsal margin of the mouth posterior to the rostral. In the ancestral condition they contact the rostral, the anterior internasals, the nasals, the postnasals, the cantholoreal series, the preoculars, the suboculars, the temporals, and the pre-auriculars. In the ancestral state the penultimate supralabial (the second from the posterior end) fails to contact the orbit (the suboculars or postoculars). Approximately 10 scales are seen in anguines and a higher number in diploglossines; the ancestral condition is therefore obscure. Tihen (1949a), probably because he viewed *Coloptychon* as primitive, illustrated 14 in his representation of the "gerrhonotine prototype."

Only in *Coloptychon* and *Gerrhonotus* are there as many as 13-14 supralabials.

The penultimate supralabial has come into contact with the orbit in *Abronia ochoterenai* (one side of both of the specimens examined), *A. matudai* (both sides of one and one side of another of the four specimens examined), *A. lythrochila* (three of the four specimens examined), *A. oaxacae* (83%), *A. chiszari*, *A. bogerti*, *A. kalaina*, *A. fuscolabialis* (which is generally considered to show the ancestral condition; Tihen, 1944, 1954), *A. deppii*, *A. vasconcelosii*, *A. aurita*, *A. salvadorensis*, and *A. montecristoi*.

The derived condition does not arise by the same means in all of these species. It is the result of the loss of the lowest anterior temporal in *Abronia deppii* (Figure 6c) (see above) and of the loss of the posteriormost supralabial in *A. oaxacae, A. ochoterenai, A. matudai, A. lythrochila, A. vasconcelosii, A. aurita, A. salvadorensis,* and *A. montecristoi* (Figure 6d,f). *Abronia kalaina* and *A. fuscolabialis* show no modification of scale elements, but the posterior margin of the mouth has shifted anteriorly half way along the posteriormost supralabial in *A. fuscolabialis* and to the next scale forward in *A. kalaina.* The expansion of the lower anterior temporal elements in *A. chiszari* and *A. bogerti* is responsible for the supralabial condition seen in those two species (Figure 6e) (see above).

Variation in the number of temporal elements contacting the supralabials, a character often used in gerrhonotine systematics (e.g. Tihen, 1954), is dependent on the variation in the temporals and supralabials described above and cannot be considered an independent character.

Mental. The single medial scale anteriormost on the lower jaw. It contacts the infralabials and the postmentals. No significant variation was observed in mental morphology.

Infralabials. The 7-12 scales making up the ventral margin of the mouth. They are ancestrally subequal in length and in contact with the mental, the postmentals, the sublabials, and the chinshields. Diploglossines and anguines differ in infralabial number, so the establishment of polarity in this character by outgroup comparison is impossible.

There are 8-10 infralabials in all gerrhonotines except *Coloptychon, Gerrhonotus,* and *Elgaria parva,* each with 11-12. The posterior infralabial is expanded to more than twice the length of the others in *Abronia lythrochila* and *A. vasconcelosii.*

Postmentals. The 1-2 scales immediately posterior to the mental. Although single postmentals are characteristic of the Anguinae and Diploglossinae, phylogenetic hypotheses among the gerrhonotines based on the assumption that this is ancestral are corroborated by no other characters. If two postmentals are considered primitive, corroboration is extensive. However, adhering to the criteria outlined above for polarizing characters, single postmentals must be considered ancestral at least in the initial stages of this analysis. Contact is with the mental, the infralabials, the sublabials, and the chinshields.

A single postmental is present in *Mesaspis viridiflava, M. moreleti, M. monticola, Abronia ochoterenai, A. matudai, A. lythrochila, A. vasconcelosii, A. aurita* (50%), *A. salvadorensis,* and *A. montecristoi.* Smith and Alvarez del Toro (1963) cited a specimen of *A. ochoterenai* with paired postmentals and Hidalgo (1983) stated that this state is sometimes present in *A. vasconcelosii,* although he examined only the two specimens seen in the present study, neither of which show it. Bogert and Porter (1967) cited two of 12 specimens of *A. oaxacae* as having single postmentals, but this was not observed here. The postmentals are reduced in size in *A. oaxacae* and *A. mixteca* (Figure 8b).

Sublabials. The 4-8 scales in paired rows immediately medial to the infralabials. They contact the infralabials, the postmentals and the chinshields. The ancestral sublabial number is unclear.

Coloptychon and *Gerrhonotus* show 7-8 sublabials (*G. lugoi* has six). All other species have 4-5 (rarely 6-7).

Chinshields. The large scales in paired rows making up most of the underside of the chin. Each row ancestrally consists of three large scales followed by a fourth

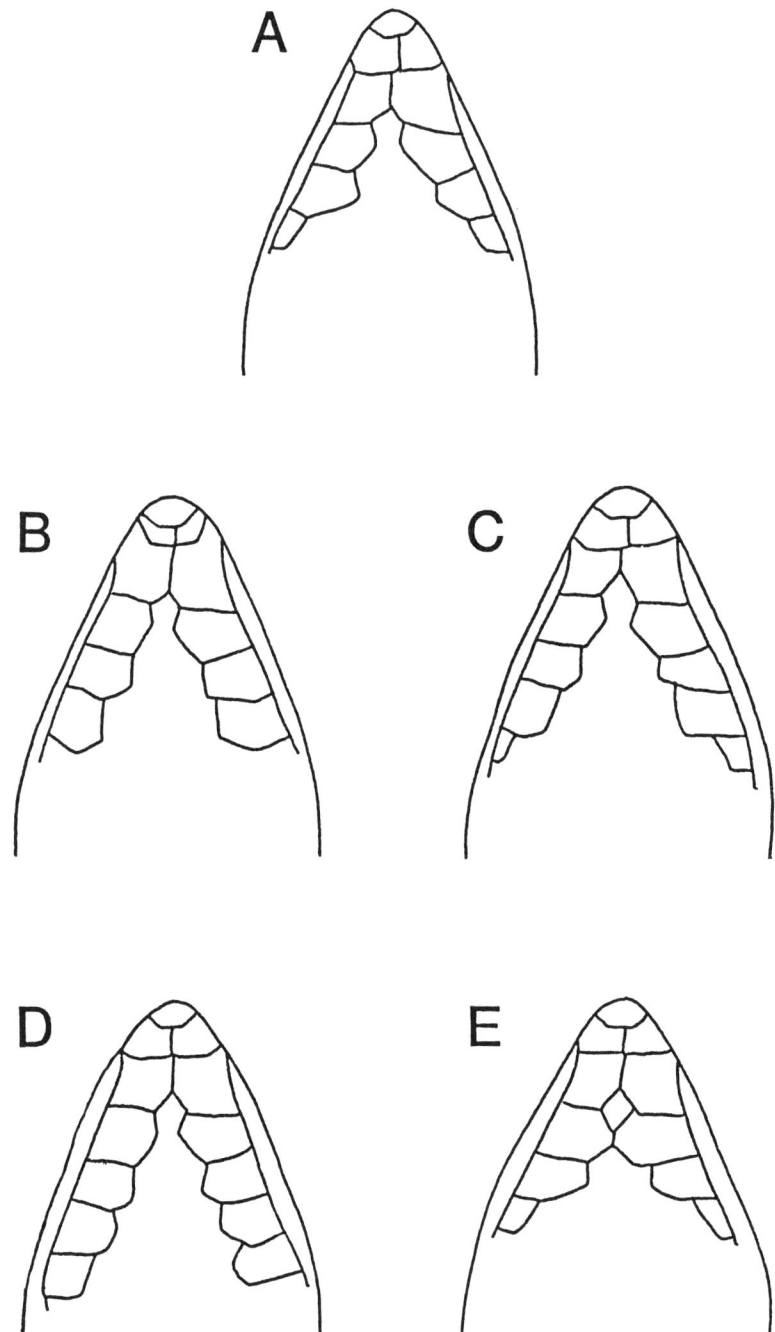

FIGURE 8. Schematic representation of chinshield variation in the Gerrhonotinae. A=general gerrhonotine variation, B=*Abronia mixteca*, C=*Gerrhonotus liocephalus*, D=*Coloptychon rhombifer*, E=*A. kalaina*.

approximately half as large as the others. Only the first pair contact each other on the midline.

In *Abronia mitchelli, A. salvadorensis, A. deppii, A. oaxacae,* and *A. mixteca,* the last chinshield has increased in size and has become subequal to the others (Figure 8b). Campbell (1982) suggested that this is also true of *A. chiszari* and *A. bogerti,* but this is not the case. In *Coloptychon rhombifer* there are five large chinshields (Figure 8d), and in *Gerrhonotus* there are four large and one smaller element (Figure 8c).

Only in *Abronia kalaina* do the second pair of shields meet on the midline (Figure 8e).

Dorsals. The scales covering the dorsum posterior to the postoccipitals and medial to the lateral folds. The dorsals are ancestrally present in many transverse (from the last postoccipital to the posterior margin of the thighs) and longitudinal (at midbody) rows. They are approximately equal to the ventral scales in size and occur in longitudinal rows parallel to the lateral fold. Keeling is ancestrally absent. At the hind limbs there are ancestrally many longitudinal rows of scales. Osteoderms are fully developed over the entire dorsum.

Most gerrhonotines possess 40-55 transverse dorsal rows. The only clear modification of this is the reduction seen in all *Abronia* to 24-39. *Barisia rudicollis* also possesses about 33 rows and *Barisia imbricata* sometimes approaches this number (Guillette and Smith, 1982).

Longitudinal rows have been reduced to 14 in *Elgaria multicarinata* (some reduced to 12 [Fitch, 1938; Smith, 1946; Stebbins, 1958] and rarely to 16 [Smith, 1946]), *E. cedrosensis, E. paucicarinata* (some with 16 [Fitch, 1938; Stebbins, 1958; Webb, 1962]), *E. panamintina, E. kingii* (rarely 16 [Taylor and Knobloch, 1940; Smith, 1946; Webb, 1962; Hardy and McDiarmid, 1969]), *Barisia rudicollis, B. imbricata imbricata* (25% with 12), *Mesaspis viridiflava,* rarely *M. juarezi* (<5%), and all *Abronia* except *A. matudai, A. mitchelli,* and *A. chiszari.* Further reduction to 10-13 is seen in *A. deppii, A. oaxacae,* and *A. mixteca.* Tihen (1954) and Gehlbach and Collette (1957) counted 14 rows in *A. oaxacae* as did Boulenger (1885) in some *A. deppii.* Martin del Campo (1939) cited 16 dorsal rows in a specimen of *A. ochoterenai* (which he referred to "*A. fimbriata*" [=*A. aurita*]), but this was not observed in the present study. Also not corroborated were the counts of 14 dorsal rows in *E. coerulea* by Boulenger (1885) and Smith (1946). *Gerrhonotus* and *M. moreleti* show an increase in longitudinal row number to 18-20. *Mesaspis moreleti* shows a general increase in scale numbers not seen in other gerrhonotines. Tihen (1949b) considered this high number primitive in his *Barisia* (*Barisia* and *Mesaspis*)/*Elgaria*/*Gerrhonotus* clade (see above), but gave no explanation for this belief.

The number of longitudinal dorsals at the hind limbs is ten in *Coloptychon, Gerrhonotus, Elgaria, Mesaspis moreleti,* and *Abronia mitchelli.* Campbell (1982) stated that *A. aurita* also has ten scales, but this was not observed in the present study. All other species have eight except *A. deppii, A. oaxacae,* and *A. mixteca* with six. The high number in *M. moreleti* is probably a result of the general increase in dorsal number described above.

Keeling is weak in *Elgaria cedrosensis, E. paucicarinata, E. panamintina, E. kingii, Mesaspis antauges, M. juarezi, viridiflava, M. monticola,* and *M. moreleti* as well as in all

Abronia, especially *A. deppii*, *A. oaxacae*, and *A. mixteca*, in which keeling is very slight. In *Mesaspis*, particularly reduced keeling is seen in *M. juarezi*, *M. antauges* and *M. viridiflava*. Keeling is also weak in some *E. coerulea*, but less so than in the species just listed. Keels are virtually absent in *G. lugoi* and *E. parva*, in which loss of keeling is probably correlated with reduction in body size, and in *C. rhombifer*. It may be that *C. rhombifer*, which retains the ancestral large size, also retains the ancestral lack of keels, while this condition is secondarily derived in the other species. Martin (1955) states that *A. taeniata* shows less keeling than does *A. graminea*. This was not apparent in this study. In *Barisia rudicollis*, keeling has been accentuated to such a degree that the dorsals have become virtually acuminate (see illustration in Tihen, 1949b).

In conjunction with their reduced number, dorsal scales are considerably larger than ventral scales in all *Abronia*. Those on the flanks are posteromedially rounded so that the longitudinal rows have become oblique relative to the lateral fold in *A. deppii*, *A. oaxacae*, and *A. mixteca* (Figure 7). Campbell (1982) described *A. mitchelli* as sharing this character, but this is a mistaken impression resulting from the position in which the only known specimen was preserved.

Osteoderm development on the dorsum is reduced in *Coloptychon rhombifer*, *Abronia kalaina*, *A. oaxacae*, and *A. mitchelli*, and absent from *A. mixteca* and *A. deppii* (see Good and Schwenk, 1985, for discussion). As there is an ontogenetic increase in osteoderm development in gerrhonotines, the condition in *A. chiszari* and *A. bogerti*, both known from single juvenile specimens, is unknown. A juvenile specimen of *A. reidi* (the paratype) was the only one available for this analysis, so the condition in that species is uncertain, although Werler and Shannon (1961) stated that it has reduced osteoderms.

Nuchals. The dorsal scales on the nape flanked laterally by the ears and the granular areas posterior to the ears. Immediately anterior are the postoccipitals. These scales ancestrally occur in many longitudinal rows (i.e., there are many scales in each transverse row).

The number of longitudinal nuchal rows is 12 in *Coloptychon rhombifer*. It is reduced to ten in *Gerrhonotus*, *Elgaria* (50% of *E. coerulea* had eight. *Elgaria kingii* rarely shows 12; Hardy and McDiarmid, 1969), and *Mesaspis moreleti*, and to eight in all other species except *Abronia* in which it is reduced to six. Exceptions in *Abronia* are seen in *A. chiszari* and *A. bogerti* which have eight scales, *A. graminea* which is often reduced to five (0.5%) or four (35%), and *A. oaxacae* with four. There is a strong geographic component to the variation seen in *A. graminea* with the population at Puerto del Aire in eastern Mexico showing predominantly four scales while those on Pico de Orizaba and Cofre de Perote show primarily six. The type of *A. mixteca* was described as having four nuchals (Bogert and Porter, 1967) and Smith and Alvarez del Toro (1962, 1963) cited a specimen of *A. lythrochila* with eight but none of the specimens of these species examined in this study showed anything other than six. Smith and Smith (1981) described the only known specimen of *A. chiszari* as having seven nuchals but this is due to the partial fusion of the two medial scales. *Barisia rudicollis* also shows a reduction in nuchal number to 4-6, but this is due to an expansion of the neck granulars in this species not seen in *Abronia*. *Mesaspis moreleti* shows ten nuchals probably as a result of the general increase in dorsal scales in that species. In the specimens of *E. coerulea* with ten nuchals, the lateral two are much reduced.

Neck scales. Neck scales occur in the region between the ear and the forelimb. They separate the nuchals from the ventrals of the throat and in the ancestral state consist of small granular scales which show an abrupt transition to the ventrals.

The neck scales show a slight increase in size in *Abronia chiszari*, *A. bogerti*, *A. kalaina*, *A. fuscolabialis*, *A. taeniata*, and *A. graminea*, and a greater increase in *A. deppii*, *A. mixteca*, and *A. oaxacae*. In these latter species the neck scales approach the ventrals in size. There is some confusion about the definition of "large" in discussions of neck scales in previous works (see Tihen, 1954). Tihen's suggestion that neck scales of *A. taeniata* may be larger than those of *A. graminea* is incorrect. Tihen (1949a) characterized *Coloptychon* as lacking granular neck scales, but it shows the ancestral condition.

A gradual transition of neck scales to ventrals occurs in *Abronia kalaina* and *A. fuscolabialis*. Expansion of the neck scale area in *Barisia rudicollis* is discussed above.

Lateral fold. The fold ancestrally running longitudinally from the ear to the hind limb. In the ancestral condition it is well developed, containing several rows of granular scales. There is an abrupt transition between these granulars and the dorsals.

The lateral fold retains the primitive unreduced state in *Gerrhonotus*, *Elgaria*, and *Barisia*. Reduction occurs in a progression from slight reduction in *Mesaspis* to moderate reduction in all *Abronia* except *A. deppii*, *A. mixteca*, and *A. oaxacae*, to extensive reduction in those three species. Reduction is also seen in *Coloptychon rhombifer*. The lateral fold has been lost in all *Abronia* between the ear and the forelimb. A more-or-less gradual transition from fold granulars to dorsals is seen in *A. kalaina*, *A. fuscolabialis*, *A. graminea*, *A. taeniata*, *A. deppii*, *A. oaxacae*, and *A. mixteca*.

Ventrals. The scales making up the ventral surface of the lizard. They are ancestrally in many transverse (from the mental to the vent) and 10 longitudinal (at midbody) rows. This latter is the condition seen in anguines; diploglossine ventrals are arranged differently and are probably not comparable with those of gerrhonotines. All scales are approximately the same width. Ten longitudinal ventral rows are ancestrally present between the forelimbs.

Twelve longitudinal ventral rows are present in all gerrhonotines except *Coloptychon rhombifer*, which retains the ancestral 10 scales, and *Gerrhonotus lugoi*, rarely *G. liocephalus* (Tihen, 1954), *Barisia rudicollis*, *Abronia lythrochila*, *A. vasconcelosii*, *A. aurita*, *A. salvadorensis*, *A. kalaina*, *A. fuscolabialis*, *A. taeniata* (30%), *A. graminea* (30%), *A. deppii* (73%), *A. oaxacae*, and *A. mixteca*, all of which have 14. Hartweg and Tihen (1946) observed 14 rows in one of four specimens of *A. ochoterenai* they examined, but it is questionable whether this specimen was really *A. ochoterenai* and not *A. lythrochila* (Smith and Alvarez del Toro, 1963). Mártin del Campo (1939) counted 12 rows in a specimen of "*A. fimbriata*" (=*A. aurita*), but Smith and Alvarez del Toro allocated this specimen to *A. ochoterenai*. Tihen (1949b) and Stebbins (1958) stated that *Mesaspis gadovii* occasionally has ten rows.

The lateral rows of ventral scales are expanded in *Abronia ochoterenai* and *A. matudai*. As a result, the number of longitudinal rows in these species is secondarily reduced to 12.

Eight longitudinal ventral rows are present between the forelimbs in *Mesaspis*. This is not correlated with any general reduction in longitudinal ventral rows.

Limb scales. Among the outgroups, only diploglossines have limbs. Ancestral gerrhonotine limb scale features include a lack of subgranulars on the leading edge of the shank and the presence of nongranular scales on the trailing edge of the limbs. The polarity of the sharpness of the transition from granular to nongranular scales on the forearm is more open to question although, a posteriori, granular scales are probably derived. At least in the early stages of this analysis, this latter character is unpolarizable.

Gerrhonotines vary in the size of the small scales on the trailing edge of the limbs. These scales are more-or-less granular in *Barisia, Mesaspis, Elgaria*, and most *Abronia*. They are larger in *Coloptychon, Gerrhonotus, A. deppii, A. mixteca*, and *A. oaxacae*; much more so in the latter three species. There is a tendency to gradual transition from granular to nongranulars on the forearm in *A. chiszari, A. bogerti, A. kalaina, A. fuscolabialis, A. taeniata, A. deppii, A. graminea, A. mixteca*, and *A. oaxacae*. Subgranular scales are present on the leading edge of the shank in *Mesaspis*.

There is a concomitant loss of keeling on the limbs in those species showing reduced dorsal keeling.

Caudal scales. The scales of the tail. These scales are arranged in whorls, with decreasing numbers of scales per whorl posteriorly. The ancestral state is a large number of whorls and 20-24 scales in the whorls near the vent.

There is much overlap in number of caudal whorls among gerrhonotine species, but there is a general trend to lower numbers in *Barisia, Mesaspis*, and *Abronia*, which usually have fewer than 100, while *Coloptychon, Gerrhonotus*, and *Elgaria* usually have more. The number of scales in the fifth whorl posterior from the vent is reduced to 15-17 in *Barisia, Mesaspis*, and *Abronia*.

There is a decrease in keeling in those species showing decreased dorsal keels (see above).

Body form. Diploglossines vary in elongation of the body. All anguines are elongate, long-tailed, and limbless. This condition in the Anguinae is almost certainly derived. However, no attempt has been made to adjust for this problem in any other character, and the ancestral gerrhonotine body form must be assumed to be elongate. If the resulting polarizations are incorrect, it should become apparent in the light of other characters. In the ancestral state, the head is not widened or depressed and the snout is relatively elongate. The limbs are relatively short and weakly clawed.

Small size has been achieved in *Gerrhonotus lugoi, Elgaria parva*, and all *Mesaspis* except *M. gadovii*. A stocky, relatively short-tailed body plan is seen in *Barisia, Mesaspis, Abronia*, and *E. coerulea* (compare Figures 7, 9, and 10, to Figures 11, 12, and 13). Long, well clawed limbs are characteristic of *Abronia* (Figure 7) and *B. rudicollis*.

A relatively widened and depressed head is seen in *Abronia* (Figure 7) The head is particularly broadened in *Abronia* species other than *A. mitchelli, A. ornelasi*, and *A. reidi*.

The snout is shortened and deepened in *Barisia* (Figure 10).

Color pattern. Color pattern variation is difficult to break down into characters, but certain trends are apparent. The groupings suggested by these characters are less clear-cut

FIGURE 9. Photograph of a 66 mm SV preserved specimen of *Mesaspis moreleti* (MVZ 104169).

FIGURE 10. Photograph of a 109 mm SV preserved specimen of *Barisia imbricata* (MVZ 196852).

FIGURE 11. Photograph of a 108 mm SV preserved specimen of *Elgaria kingii* (MVZ 191078).

FIGURE 12. Photograph of a 154 mm SV preserved specimen of *Gerrhonotus liocephalus* (MVZ 128079).

than those suggested by scale characters, and certain character states, such as the degree of sexual dichromatism, are unknown for several forms because of lack of material.

The ancestral adult color pattern by outgroup comparison is a broad dorsal stripe, probably of some shade of brown, flanked by darker sides (Figure 9). The venter and head are unmarked. Sexes are similar.

If cross-banding in all gerrhonotines showing it is considered homologous, ambiguous results occur: it would be equally parsimonious to claim that cross banding is ancestral (not ancestral for the gerrhonotine lineage, but arising in it before the divergence of the modern genera) and that a dorsal stripe is ancestral. The distinctive dorsal patterning in each of the banded gerrhonotine genera leads me to accept the latter hypothesis; substantial dorsal cross banding has arisen probably independently in *Coloptychon, Gerrhonotus, Elgaria,* and *Abronia*. However, no strong arguments can be made one way or the other. The condition in *Coloptychon* is one of large light rhomboidal patches on a darker background (Figure 13). *Gerrhonotus* shows mostly narrow light bands (Figure 12). *Elgaria* and *Abronia* both have dark cross bands on a lighter background, but *Elgaria* (Figure 11) usually has more and narrower bands than does *Abronia* (Figure 7).

Elgaria kingii, E. panamintina, and *E. parva* share a pattern of rather suffuse dark pigmentation between the darker cross bands (Figure 11). In *Abronia* there is a dichotomy between those species with 6-8 dorsal cross bands (*A. salvadorensis, A. graminea, A. taeniata, A. deppii, A. oaxacae,* and *A. mixteca*) and those with 9-11 (all other species). *Abronia reidi* and *A. ornelasi* share a similar light coloration on the posterior edges of their dorsal scales. In none of these latter three cases can polarization be determined, although, in the light of relationships suggested by other characters, the *E. kingii/E. panamintina/E. parva,* 6-8 banded *Abronia,* and *A. reidi/A. ornelasi* patterns probably show the derived conditions.

FIGURE 13. Photograph of a 120 mm SV preserved specimen of *Coloptychon rhombifer* (UCR 3143).

In *Barisia imbricata* and *B. levicollis*, there is a tendency to an almost complete lack of dorsal patterning, including banding on the sides (Figure 10). A similar condition occurs in the bright green *Abronia* (*A. ochoterenai*, *A. matudai*, and *A. graminea*).

Ventral markings occur in five types. In *Coloptychon*, chevron-like dark bands are seen on the throat and at least on the anterior part of the trunk. Longitudinal ventral stripes are seen in *Elgaria*, between the scale rows in *E. coerulea* and on the scales in the other species (which of these is the derived state is not discernible). In *E. kingii* these stripes are broken into an irregular spotting pattern, and in *E. parva* ventral markings are absent. The third type of ventral marking is the characteristic speckling seen in *Mesaspis*. In *Abronia kalaina* and *A. fuscolabialis*, cross bands are present on the posterior part of the trunk. *Abronia chiszari* and *A. bogerti* share the fifth form of ventral coloration, the presence of scattered light speckling.

Black and white markings on the lateral fold, caused by a continuation of the flank bars into the fold, are seen in *Elgaria kingii*, *E. panamintina*, *E. cedrosensis*, and *E. paucicarinata*.

Head markings occur primarily in *Elgaria*. These include more-or-less random spotting patterns seen in most species (this is also seen in many *Mesaspis moreleti*) and characteristic black and white markings on the labials in *E. kingii*, *E. cedrosensis*, and *E. paucicarinata*. Characteristic pale labial markings of a different sort are seen in *Mesaspis*.

Sexual dichromatism is characteristic of many species of *Barisia*, *Mesaspis*, and *Abronia*.

PHYLOGENETIC ANALYSIS

The Gerrhonotinae comprises six almost certainly monophyletic units which are here referred to the genera *Coloptychon*, *Gerrhonotus*, *Elgaria*, *Barisia*, *Mesaspis*, and *Abronia*. Each of these genera possesses a suite of derived external features unique in the subfamily (see diagnoses below), and I have elsewhere (Good, 1987b) provided further derived features of cranial osteology substantiating the monophyly of all but *Coloptychon*. This also is borne out by a PAUP analysis of all 38 species combined. In order to simplify the data matrices and facilitate discussion of character state changes and relationships among the gerrhonotine species, the following discussion proceeds from the assumption of generic monophyly and will therefore deal with separate analyses among the six genera, and within each of them.

RELATIONSHIPS AMONG THE GERRHONOTINE GENERA

The distribution of character states among the six gerrhonotine genera in the characters discussed in the last chapter (and summarized in Appendix B) is given in Appendix C. Some characters are autapomorphies of single genera or of subsets of single genera and hence provide no evidence in an intergeneric phylogenetic analysis. Characters that are variable within genera are uninformative about the states ancestral for those genera because no intrageneric phylogenies are available at this stage in the analysis; any character state seen in a genus is as likely as any other to be ancestral for that genus. For the purposes of discerning relationships, such characters were coded as unknown for the variable genera. Following this procedure, only characters showing derived states fixed in two or more genera were useful; these synapomorphies are listed in Table 1. Also included in the table are unpolarizable characters which nevertheless contain potentially useful phylogenetic information that can be mapped a posteriori onto the phylogeny suggested by polarizable characters. Given the character state distributions in Table 1, a single hypothesis for the phylogenetic relationships among the gerrhonotine genera is unambiguously suggested (Figure 14), and this hypothesis is in full agreement with the results of my study of cranial characters (Good, 1987b). *Coloptychon*, a monotypic, poorly known genus from Costa Rica and Panama, is the sister taxon to all other gerrhonotines. This is unambiguously

TABLE 1. Distribution of derived and ancestral states of potentially phylogenetically informative characters among the genera of gerrhonotine lizards

Character	*Coloptychon*	*Gerrhonotus*	*Elgaria*	*Barisia*	*Mesaspis*	*Abronia*
17-1	0	0/1	1	1	1	1
34	a	a/b	b	b	b	b
37	b	b	a	a	a	a
40	0	0/1	1	1	1	1
52	0	0/1	1	1	1	0/1
60	b	b	a	a	a	a
63	a	a	a/b	b	b	b
67	a	a	b	b	b	b
73-1	0	1	1	1	1	1
76-1	0	1	1	1	1	1
76-2	0	0	0	1	0/1	1
79-1	1	0	0	0	1	1
79-2	0	0	0	0	1	1
82-1	0	1	1	1	1	1
85	0	0	1	1	1	0/1
88	a	a	a	b	b	b
89	0	0	0	1	1	1
90	0	0	0/1	1	1	1
105	0	0	0	1	1	1

Note: Characters are designated by numbers corresponding to those in Appendix B. Some of the general characters in Appendix B are subdivided into multiple characters (see Appendix C). Polarity is established by outgroup comparison with the Diploglossinae and Anguinae. When polarity is ambiguous, letter designations are given to alternative states. 0=ancestral, 1=derived.

suggested by three synapomorphies of the non-*Coloptychon* gerrhonotines: acquisition of keeling (character 73-1), a reduction in nuchal number (character 76-1), and an increase in longitudinal ventral number (character 82-1). Four more characters may also be synapomorphies of this non-*Coloptychon* clade (with a reversal in some *Gerrhonotus*), but a variable condition in *Gerrhonotus* makes it equally likely that they are synapomorphies rather of a clade containing *Elgaria*, *Barisia*, *Mesaspis*, and *Abronia*, with a parallelism in some *Gerrhonotus*: a reduction in canthal/loreal number (character 17-1), a decrease in preocular number (character 34; this character is unordered and placed on the phylogram a posteriori), a decrease in the number of temporals per row (character 40), and a decrease in the number of primary temporals contacting the orbit (character 52).

In addition to these four characters which might be synapomorphies of *Elgaria*, *Barisia*, *Mesaspis*, and *Abronia*, four others unambiguously agree on suggesting this alliance, although only one of them is polarizable a priori (character 85): differentiation of the suboculars from the preoculars and postoculars (character 37), a decrease in supralabial number (character 60), a decrease in sublabial number (character 67), and acquisition of

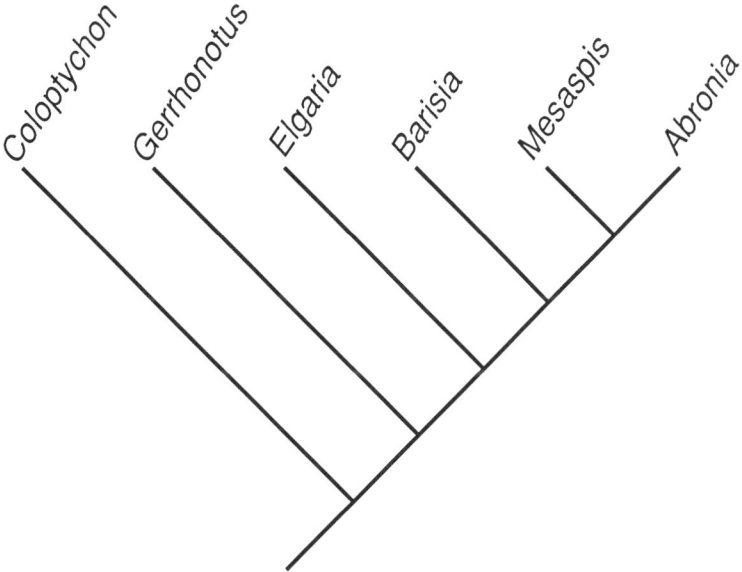

FIGURE 14. Phylogenetic relationships among the gerrhonotine genera as suggested by the analysis of external characters. Character state changes are discussed on page 35.

granular scales on the trailing edges of the limbs (character 85). There are, in addition, two characters that are variable in *Elgaria*, and hence could just as easily be synapomorphies of this clade (with a reversal in some *Elgaria*), or a synapomorphy of *Barisia*, *Mesaspis*, and *Abronia* (seen in parallel in some *Elgaria*): an increase in infralabial number (character 63, polarized a posteriori) and acquisition of a shortened, stocky body form (character 90).

The montane Middle American clade of *Barisia*, *Mesaspis*, and *Abronia* is also diagnosed by four unambiguous characters: a reduction in longitudinal nuchal scale rows (character 76-2), a decrease in caudal scale row number (character 88, polarized a posteriori), a decrease in the number of scales per caudal whorl (character 89), and sexual dichromatism (character 105).

Mesaspis and *Abronia* show a synapomorphy in reduction in the lateral fold (character 79), a feature seen in parallel in *Coloptychon* (79-1) but not to the same degree as in *Mesaspis* or *Abronia* (79-2). The alliance of *Mesaspis* and *Abronia* with *Coloptychon* suggested by this character is outweighed easily by the 17 characters discussed above suggesting alliances with other groups of gerrhonotines.

Aside from the homoplastic events discussed above, three others are required by these results: the number of primary temporals (character 52) and the size of the scales on the trailing edge of the limbs (character 85) increase secondarily in some *Abronia* and the number of nuchals (character 76-2) increases secondarily in some *Mesaspis*.

Given this hypothesis of phylogenetic relationships among gerrhonotine genera, it is possible to use the other genera as sequential outgroups in polarizing characters for intrageneric analyses, using "global parsimony" as discussed by Maddison et al. (1984). For instance, for analysis within *Elgaria*, the *Barisia-Mesaspis-Abronia* clade is the immediate outgroup, the outgroup to this clade is *Gerrhonotus*, and the outgroup to all of these genera is *Coloptychon*.

Intrageneric relationships within *Coloptychon* and *Gerrhonotus* need no analysis; the former is monotypic, and the latter contains only two species as currently recognized.

RELATIONSHIPS WITHIN *ELGARIA*

Scale characters among *Elgaria* species (Table 2) are quite invariant relative to intrageneric variation in other gerrhonotine genera. However, sufficient characters remain to suggest a reasonable phylogenetic hypothesis (Figure 15). *Elgaria coerulea* appears to be the sister species to all other *Elgaria*. This is suggested by two synapomorphies of the clade containing all other species: a reduction in longitudinal dorsals (character 71-2) and the acquisition of longitudinal dark stripes lying on the ventral scale rows (character 100-3). A further possible synapomorphy of this clade is a long slender body form (character 90), but it is impossible on the basis of outgroup comparison to determine whether the *E. coerulea* condition (relatively short and stocky) or the condition in the clade containing all other *Elgaria* is the derived one.

Within the non-*Elgaria coerulea* clade, *E. multicarinata* is the sister species to all others, as suggested by three characters allying the remaining five species: a somewhat triangularly shaped lower subocular (character 39), reduced keeling (character 73-2), and black and white markings on the lateral fold (character 101). It is impossible to determine, on the basis of this analysis, the relationships among the three lineages in the remainder of *Elgaria*: that containing *E. cedrosensis*, that containing *E. paucicarinata*, and that containing *E. panamintina*, *E. kingii*, and *E. parva*. This latter clade is diagnosed, however, by the presence of suffuse, dark pigmentation between the dorsal crossbands (character 96). The presence of black and white markings on the labials (character 103) is either a synapomorphy of this entire clade, with a secondary loss in *E. panamintina*, or it evolved in parallel in the common ancestor of *E. cedrosensis* and *E. paucicarinata* and the common ancestor of *E. kingii* and *E. parva*. If the latter is the case, the relationships of *Elgaria* can be further resolved; however, because either scenario is equally likely, resolution on the basis of morphology is not possible. Biogeographic implications, however, suggest that the former may be more likely (see section on biogeography below).

Elgaria parva is the most enigmatic of the *Elgaria* species with regard to its phylogenetic placement, showing the ancestral condition in lacking a triangular lower subocular (character 39) and ventral markings (character 100-3). These characters would combine to suggest that *E. parva* should be placed outside of all of the other *Elgaria* with the possible exception of *E. coerulea*. However, there are six characters (see above) which suggest that it is imbedded well within the non-*E. coerulea* clade; characters 39 and 100-3 are probably reversals.

TABLE 2. Distribution of derived and ancestral states of potentially phylogenetically informative characters among the species of *Elgaria*

Character	coer	multi	pauci	cedro	pana	king	parv
39	0	0	1	1	1	1	0
71-2	0	1	1	1	1	1	1
73-2	0	0	1	1	1	1	1
90	a	b	b	b	b	b	b
96	0	0	0	0	1	1	1
100-3	0	1	1	1	1	1	0
101	0	0	1	1	1	1	1
103	0	0	1	1	0	1	1

Note: Characters are designated by numbers corresponding to those in Appendix B. Some of the general characters in Appendix B are subdivided into multiple characters (see Appendix C). Polarity is established by outgroup comparison with the clade containing *Barisia*, *Mesaspis*, and *Abronia* as the first, *Gerrhonotus* as the second, and *Coloptychon* as the third sequential outgroups. When polarity is ambiguous, letter designations are given to alternative states. 0=ancestral, 1=derived. coer=*E. coerulea*, multi=*E. multicarinata*, pauci=*E. paucicarinata*, cedro=*E. cedrosensis*, pana=*E. panamintina*, king=*E. kingii*, parv=*E. parva*.

RELATIONSHIPS WITHIN *BARISIA*

Barisia contains only three species, so phylogenetic analysis is straightforward, there being only three dichotomously branching dendrograms possible. No characters (Table 3) link *B. rudicollis* with *B. levicollis*. Two link *B. rudicollis* with *B. imbricata*: a decrease in transverse dorsal row number (character 70) and in longitudinal dorsal row number (character 71-2). Four characters unite to suggest, on the other hand, that *B. imbricata* and *B. levicollis* are sister species: postnasal-anterior loreal fusion (character 13), a sunken appearance of the frontal (character 29), an increase in the number of temporals per row (character 40), and a loss of dorsal patterning in the adult (character 99). There are therefore two characters suggesting one pattern and four characters suggesting another. Parsimony dictates acceptance of the latter; *B. imbricata* and *B. levicollis* are probably sister species (Figure 16).

RELATIONSHIPS WITHIN *MESASPIS*

Four synapomorphies (Table 4) of the other five species combine to suggest that *Mesaspis gadovii* is the sister species to all other *Mesaspis* (Figure 17): these are supranasal expansion (character 10-1), high canthal/loreal variability (character 22; the state in *M. antauges* is inferred a posteriori), reduction in keeling (character 73-2), and reduced body size (character 91).

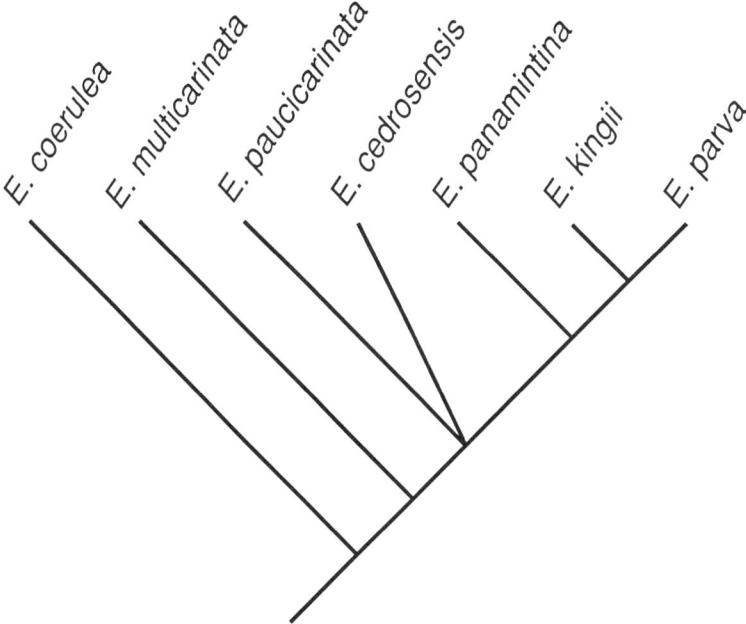

FIGURE 15. Phylogenetic relationships among the species of *Elgaria* as suggested by the analysis of external characters. Character state changes are discussed on page 38.

Mesaspis antauges and *M. juarezi* are united by sharing three derived characters: the occasional presence of a postrostral (character 3), a broad contact between the frontal and the interparietal (character 27), and an elongate anterior superciliary (character 33).

Four characters combine to unite *Mesaspis moreleti* and *M. monticola*: reduction of the posterior internasals (character 7-2), expansion of the supranasals to the midline (character 10-2; seen in parallel in some *M. juarezi*), division of the cantholoreal into two elements (character 17-2), and contact of the anterior loreal and the preocular (character 20). Opposed to the sister species status of *M. moreleti* and *M. monticola* is a single character linking *M. viridiflava* and *M. monticola*: postnasal-anterior loreal fusion (character 13). This single character is clearly outweighed by the four linking *M. moreleti* and *M. monticola*, and probably represents either a parallel acquisition of this fusion in *M. viridiflava* and *M. monticola* or a synapomorphy of the *moreleti* group that is secondarily lost in *M. moreleti*. Either alternative is equally parsimonious.

The absence of a frontonasal (character 15) suggests that *Mesaspis viridiflava* is allied with *M. antauges* and *M. juarezi* and the presence of a single postmental (character 65) suggests that it is allied with *M. moreleti* and *M. monticola*. Frontonasal absence is a highly plastic character in the Gerrhonotinae, both inter- and intraspecifically, while postmental condition appears to be fairly conservative. For this reason, and because *M. viridiflava* also shares postnasal-anterior loreal fusion (character 13) with *M. monticola* but no other characters only with *M. antauges* or *M. juarezi*, I consider *M. viridiflava* more likely to be close to *M. moreleti* and *M. monticola*.

TABLE 3. Distribution of derived and ancestral states of potentially phylogenetically informative characters among the species of *Barisia*

Character	*rudicollis*	*imbricata*	*levicollis*
13	0	1	1
29	0	1	1
40	0	0/1	0/1
70	1	1	0
71-2	1	1	0
99	0	1	1

Note: Characters are designated by numbers corresponding to those in Appendix B. Some of the general characters in Appendix B are subdivided into multiple characters (see Appendix C). Polarity is established by outgroup comparison with the clade containing *Mesaspis* and *Abronia* as the first, *Elgaria* as the second, *Gerrhonotus* as the third, and *Coloptychon* as the fourth sequential outgroups. 0=ancestral, 1=derived. Species are referred to by their specific epithets.

Two further characters cannot be mapped onto the phylogram in Figure 17 unambiguously. First, contact of the frontal with the frontonasal (character 16) is seen in *Mesaspis gadovii*, *M. moreleti*, and *M. monticola*. However, because the frontonasal is lost in *M. viridiflava*, the state present in its pre-loss ancestor is unknown. Thus, this character could be mapped onto the phylogram in Figure 17 in a variety of different, equally parsimonious ways. Second, reduction in the number of lateral supraoculars (character 30-1) is seen in the clade containing *M. viridiflava*, *M. moreleti*, and *M. monticola* as well as in *M. juarezi*. This is equally likely to be a parallel acquisition in these two clades or a synapomorphy of all *Mesaspis* excluding *M. gadovii*, with a subsequent reversal in *M. antauges*. Neither of these characters suggests any pattern strongly enough to counterbalance the evidence suggested by those discussed above.

RELATIONSHIPS WITHIN *ABRONIA*

Because of the large number of species (19) involved in an analysis of relationships within *Abronia*, homoplasies are common (Table 5). In spite of this, a single phylogram is clearly suggested (Figure 18): *Abronia* appears to be divisible into four monophyletic groups. The *mitchelli* group is composed solely of *A. mitchelli*. The *reidi* group contains *A. reidi* and *A. ornelasi* and is diagnosed by expansion of the supranasals to the midline (character 10-2), an increase in the number of primary temporals contacting the orbit (character 52), characteristic light coloration on the posterior edges of the scales (character 98), and contact of the frontal and frontonasal (character 16). This last character is seen in parallel in some *A. aurita*, *A. graminea*, and *A. taeniata*. The *aurita* group, containing *A. montecristoi*, *A. salvadorensis*, *A. ochoterenai*, *A. lythrochila*, *A. matudai*, *A. aurita*, and *A. vasconcelosii*, is diagnosed primarily by the presence of a single postmental (character 65). Two other characters diagnose the group but are seen in parallel elsewhere in single *Abronia* species:

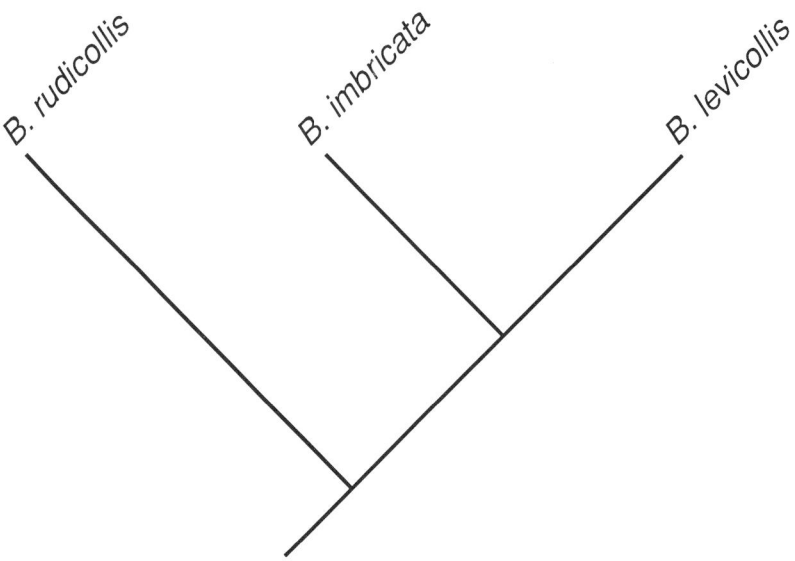

FIG. 16. Phylogenetic relationships among the species of *Barisia* as suggested by the analysis of external characters. Character state changes are discussed on page 39.

three suboculars (character 36-1) are seen also in some *A. ornelasi* (*reidi* group) and loss of the posterior supralabials (character 61) is seen also in *A. oaxacae* (*deppii* group). The *deppii* group is diagnosed by two characters: an increase in granular neck scale size (character 77-1) and a gradual transition from granulars to non-granulars on the limbs (character 86), both of which are unique among gerrhonotines.

The *aurita* and *deppii* groups are sister taxa sharing four synapomorphies: loss of subocular-temporal contact (character 38), reduction of the fourth temporal row (character 42-1), reduction in secondary temporal number (character 46), and a widened and depressed head (character 93-2). Relationships among the *aurita*/*deppii*, *mitchelli*, and *reidi* groups, however, are less clear. A single character, supranasal expansion (character 10-1, seen in parallel in *A. matudai* in the *aurita* group), suggests that the *mitchelli* and *reidi* groups are sister taxa. Another, a reduction in longitudinal dorsal number (character 71-2, secondarily lost in *A. chiszari* in the *deppii* group), suggests an alliance of the *reidi* and *aurita*/*deppii* groups.

Within the *aurita* group, *Abronia montecristoi* is probably the sister species to all others. These remaining species possess a single synapomorphy, an increase in longitudinal ventral scale rows (character 82-2). This, in parallel, is also a synapomorphy of several *deppii* group species (see below). Two characters suggest that, among the remaining six *aurita* group species, *A. salvadorensis* is the sister species to the others. The synapomorphies of the five remaining *aurita* group species are protuberant supra- auriculars (character 58) and granular pre-auriculars (character 59), this latter also being in parallel a synapomorphy of *A. graminea* and *A. taeniata* in the *deppii* group (see below). Two lineages are apparent within the remainder of the *aurita* group: one containing *A. matudai* and *A. ochoterenai*, diagnosed by the expansion of the lateral longitudinal ventral rows (character 83) and a loss

TABLE 4. Distribution of derived and ancestral states of potentially phylogenetically informative characters among the species of *Mesaspis*

Character	*gadovii*	*antauges*	*juarezi*	*viridiflava*	*moreleti*	*monticola*
3	0	0/1	0/1	0	0	0
7-2	0	0	0	0	0/1	1
10-1	0	1	1	1	1	1
10-2	0	0	0/1	0	0/1	1
13	0	0	0	1	0	1
15	0	0/1	0/1	1	0	0
16	0/1	0	0/1	N	1	0/1
17-2	0	0	0	0	0/1	0/1
20	0	0	0	0	0/1	0/1
22	0	N	1	1	1	1
27	0	1	1	0	0	0
30-1	0	0	0/1	1	0/1	1
33	0	1	1	0	0	0
65	0	0	0	1	1	1
73-2	0	1	1	1	1	1
91	0	1	1	1	1	1

Note: Characters are designated by numbers corresponding to those in Appendix B. Some of the general characters in Appendix B are subdivided into multiple characters (see Appendix C). Polarity is established by outgroup comparison with *Abronia*, *Barisia*, *Elgaria*, *Gerrhonotus*, and *Coloptychon* as sequential outgroups. 0=ancestral, 1=derived, N = not applicable. Species are referred to by their specific epithets.

of dorsal patterning in the adult (character 99, seen in parallel in *A. mitchelli* and *A. graminea*); and the other containing *A. lythrochila*, *A. vasconcelosii*, and *A. aurita*, diagnosed by loss of the third primary temporal (character 49) and loss of the fourth temporal row (character 42-2), both characters seen in parallel in *A. oaxacae* and *A. mixteca* in the *deppii* group. Within this latter clade, the sister species status of *A. lythrochila* and *A. vasconcelosii* is suggested by an expansion of the posteriormost infralabial (character 64).

The *deppii* group consists of four distinct clades. One of these, containing *Abronia chiszari* and *A. bogerti*, is diagnosed by five characters: an elongate anterior superciliary (character 33), reduction in the third temporal row (character 43), expansion of the lower primary temporals (character 51), characteristic scattered light ventral flecking (character 100-7), and probably an increase in nuchal number (character 76-3), although this latter requires that the polarity established using the other gerrhonotine genera as outgroups be reversed. The second *deppii* group clade contains *A. fuscolabialis* and *A. kalaina* and is diagnosed by a gradual transition of neck granulars to ventrals (character 78) and characteristic cross-banding on the posterior part of the venter and the tail (character 100-6). The clade containing *A. taeniata* and *A. graminea* is diagnosed only by a reduction in pre-auricular size (character 59), a character shared in parallel with many *aurita* group

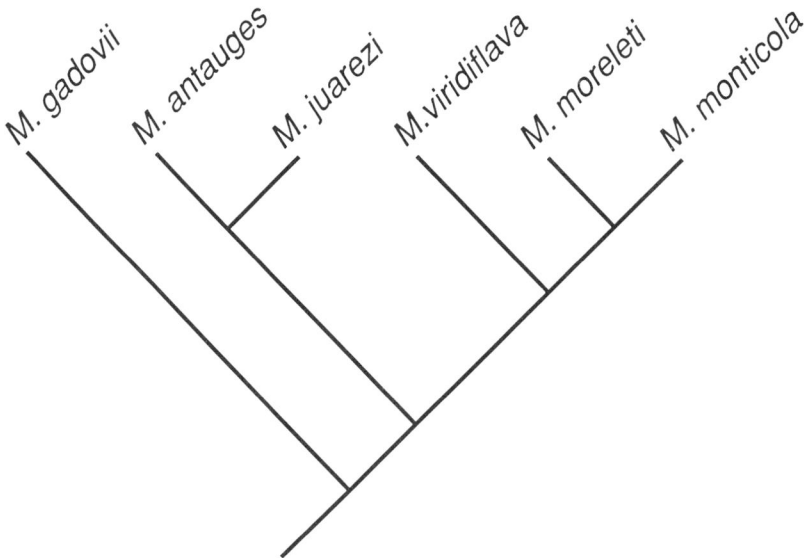

FIGURE 17. Phylogenetic relationships among the species of *Mesaspis* as suggested by the analysis of external characters. Character state changes are discussed on page 39.

species (see above). Perhaps the most strongly supported clade in all of *Abronia* is the fourth *deppii* group clade, containing *A. deppii*, *A. oaxacae*, and *A. mixteca*. Ten characters combine to suggest this grouping: pronounced knob-like head scales (character 57-2), a decrease in longitudinal dorsal scale number (character 71-3), a decrease in the number of longitudinal dorsals at the hind limb (character 72-2), strong reduction in keeling (character 73-2), posteromedially rounded flank scales (character 74), an increase in lateral neck scale size (character 77-2), a strongly reduced lateral fold (character 79-3), an increase in the size of the scales on the trailing edges of the limbs (character 85), four large chinshields (character 68-1, shared in parallel with *A. mitchelli* and *A. salvadorensis*), and reduction in dorsal osteoderms (character 75-1, shared in parallel with *A. mitchelli*, *A. reidi*, and *A. kalaina*).

Among the species *Abronia deppii*, *A. oaxacae*, and *A. mixteca*, a single character (loss of superciliary-cantholoreal contact, character 32) shared in parallel with *A. salvadorensis* suggests an alliance of *A. deppii* with *A. oaxacae*. An increase in the number of postoccipital rows (character 55-2) and loss of dorsal osteoderms (character 75-2) suggest an alliance of *A. deppii* with *A. mixteca*. However, the best hypothesis for the relationships of these three species, that *A. oaxacae* and *A. mixteca* are sister species, is suggested by three characters: three interoccipitals (character 54-2), a reduction in postmental size (character 66), and loss of the fourth temporal row (character 42-2), this last being found in parallel also as a synapomorphy of three *aurita* group species (see above).

Relationships among the four clades within the *deppii* group are probably as follows: the *Abronia taeniata*/*A. graminea* and *A. deppii*/*A. oaxacae*/*A. mixteca* clades are sister groups and share four synapomorphies: nasal-third supralabial contact (character 4),

TABLE 5. Distribution of derived and ancestral states of potentially phylogenetically informative characters among the species of *Abronia*

Char	mi	or	re	oc	ma	au	ly	va	sa	mo	ch	bo	ka	fu	gr	ta	de	oa	mx
4	0	0	0	0	0	0	0	0	0	0	0	0	0	0	0/1	0/1	1	0/1	0/1
7-1	0	0	0	0	0	0	0	0	0	0	0	0	0	0	1	1	1	1	1
10-1	1	1	1	0	0/1	0	0	0	0	0	0	0	0	0	0	0	0	0	0
10-2	0	1	1	0	0	0	0	0	0	0	0	0	0	0	0	0	0	0	0
16	0	1	0/1	0	0	0/1	0	0	0	0	0/1	0	0	0	0/1	0/1	0/1	0	0
19	0	0	1	0	0	0	1	0	0	0	0	0	0	0	0	0	1	1	1
23	0	0	1	0	0	0	0	0	1	0	0/1	0	0	0	0	0	0	0	0
32	0	0	0	0	0	0	0	0	0	0	0	0	0	0	0	0	0/1	1	0
33	0	0	0	0	0	0	0	0	0	0	1	1	0	0	0	0	0	0	0
36-1	0/1	0	0	0/1	1	1	1	1	1	0/1	0	0	0	0	0/1	0	0	0	0
38	0	0	0	1	1	1	1	1	1	1	1	1	0/1	1	1	1	1	1	1
42-1	0	0	0	1	1	1	1	1	1	1	1	1	1	0	1	1	1	1	1
42-2	0	0	0	0	0	1	1	1	0	0	0	1	0	0	0	0	0	0	0/1
43	0	0	0	0	0	0	0	1	0	0	1	1	1	0	1	0	1	1	0
46	0	0	0	1	0	1	1	1	1	0/1	1	1	1	0	1	1	1	0	0/1
48	0	0	1	0	1	0	0	0	1	0/1	0	0	0	0	0	0	0	0	0
49	0	0	0	0	0	0/1	1	1	0	0	0	1	0	0	0	0	0	1	0/1
51	0	0	0	0	0	0	0	0	0	0	1	1	0	0	0	0	0	0	0
52	0	1	1	0	0	0	0	0	0	0	0	0	0	0	0	0	0	0	0
54-2	0	0	0	0	0	0	0	0	0	0	0	0	0	0	0	0	0	1	1
55-2	0	0	0	0	0	0	0	0	0	0	0	0	1	1	1	0	1	0	1
56-1	0	0	0	0	0	0	0	0	0	0	0	0	0	0	1	1	1	1	1
57-1	0	0	0	0	0	0	0	0	0	0	0	0	0	0	1	1	1	1	1
57-2	0	0	0	0	0	0	0	0	0	0	0	0	0	0	0	0	1	1	1

TABLE 5 (continued)

Char	mi	or	re	oc	ma	au	ly	va	sa	mo	ch	bo	ka	fu	gr	ta	de	oa	mx
58	0	0	0	1	1	1	1	1	0	0	0	0	0	0	0	0	0	0	0
59	0	0	0	1	1	1	1	1	0	0	0	0	0	0	0/1	0/1	0	0	0
61	0	0	0	0/1	1	1	0/1	1	1	1	0	0	1	1	0	0	0	0/1	0
62	0	0	0	0	0	0	0	0	0	0	0	0	0	0	0	0	0	0	0
64	0	0	0	0	0	0	0	1	1	1	0	0	0	0	0	0	0	0	0
65	0	0	0	1	1	0/1	1	1	1	1	0	0	0	0	0	0	0	0	0
66	0	0	0	0	0	0	0	0	0	0	0	0	0	0	0	0	1	1	1
68-1	1	0	0	0	0	0	0	0	1	1	0	1	1	1	1	1	1	1	1
71-2	0	1	1	1	0	1	1	1	1	0	0	0	0	0	0	0	1	1	1
71-3	0	0	0	0	0	0	0	0	0	1	0	0	0	1	0	0	1	1	1
72-2	0	0	0	0	0	0	0	0	0	0	0	0	0	0	0	0	1	1	1
73-2	0	0	0	0	0	0	0	0	0	0	0	0	0	0	0	0	1	1	1
74	0	0	0	0	0	0	0	0	0	0	0	0	0	0	0	0	1	1	1
75-1	1	0	1	0	0	0	0	0	0	0	N	N	1	0	0	0	1	0	1
75-2	0	0	0	0	0	0	0	0	0	0	N	N	0	0	0	0	1	1	1
76-3	1	1	1	1	1	1	1	1	1	1	0	0	1	1	0/1	1	0	0	0
76-4	0	0	0	0	0	0	0	0	0	0	0	0	0	0	1	0	0	1	1
77-1	0	0	0	0	0	0	0	0	0	0	1	1	1	1	1	1	1	1	1
77-2	0	0	0	0	0	0	0	0	0	0	0	0	0	0	0	0	1	1	1
78	0	0	0	0	0	0	0	0	0	0	0	0	0	0	0	0	0	0	0
79-3	0	0	0	0	0	0	0	0	0	0	0	0	1	1	0	0	1	1	1
81	0	0	0	0	0	1	0	1	1	0	0	0	1	1	1	1	1	1	1
82-2	0	0	0	1	1	1	1	1	0	0	0	0	0	0	0/1	0/1	0/1	1	1
83	0	0	0	1	1	0	0	0	0	0	0	0	0	0	0	0	0	0	0

TABLE 5 (continued)

Char	mi	or	re	oc	ma	au	ly	va	sa	mo	ch	bo	ka	fu	gr	ta	de	oa	mx
85	0	0	0	0	0	0	0	0	0	0	0	0	0	0	0	0	1	1	1
86	0	0	0	0	0	0	0	0	0	0	1	1	1	1	1	1	1	1	1
93-2	0	0	1	1	1	1	1	1	1	1	1	1	1	1	1	1	1	1	1
97	N	0	0	0	0	0	0	0	0	0	0	0	0	0	1	1	1	1	1
98	0	1	1	0	0	0	0	1	0	0	0	0	0	0	0	0	0	0	0
99	1	0	0	1	1	0	0	0	0	0	0	0	1	1	0	0	0	0	0
100-7	0	0	0	0	0	0	0	0	0	0	0	1	0	1	0	0	0	0	0
100-8	0	0	0	0	0	0	0	0	0	0	1	1	0	0	0	0	0	0	0

Note: Characters are designated by numbers corresponding to those in Appendix B. Some of the general characters in Appendix B are subdivided into multiple characters (see Appendix C). Polarity is established by outgroup comparison with *Mesaspis*, *Barisia*, *Elgaria*, *Gerrhonotus*, and *Coloptychon* as sequential outgroups. 0=ancestral, 1=derived, N = not applicable. Species are referred to by their specific epithets. mi=*A. mitchelli*, or=*A. ornelasi*, re=*A. reidi*, oc=*A. ochoterenai*, ma=*A. matudai*, au=*A. aurita*, ly=*A. lythrochila*, va=*A. vasconcelosii*, sa=*A. salvadorensis*, mo=*A. montecristoi*, ch=*A. chiszari*, bo=*A. bogerti*, ka=*A. kalaina*, fu=*A. fuscolabialis*, gr=*A. graminea*, ta=*A. taeniata*, de=*A. deppii*, oa=*A. oaxacae*, mx=*A. mixteca*.

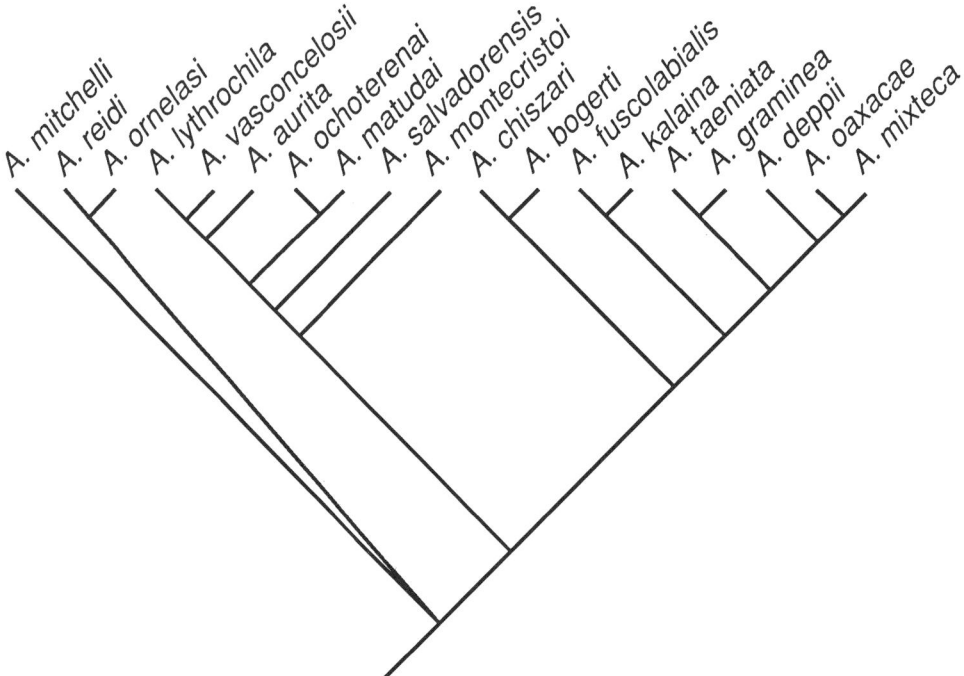

FIGURE 18. Phylogenetic relationships among the species of *Abronia* as suggested by the analysis of external characters. Character state changes are discussed on page 41.

rugose postoccipital scales (character 56-1), anterior canthal-posterior internasal fusion (character 19, seen in parallel in *A. reidi* and *A. lythrochila*), and a reduction in the number of dorsal crossbands (character 97, seen in parallel in *A. salvadorensis*). The sister group to the clade containing all five of these species appears to be the clade containing *A. fuscolabialis* and *A. kalaina*, as suggested by four characters: an increase in posterior internasal size (character 7-1), knob-like posterior head scales (character 57-1), a gradual transition of lateral fold granulars to dorsals (character 81), and a decrease in longitudinal ventral number (character 82-2, found in parallel in much of the *aurita* group). The *A. bogerti/A. chiszari* clade is then left as the sister group to all other *deppii* group species.

Figures 14-18 combine to represent the most likely hypothesis for the relationships among gerrhonotine lizards. Although further data may suggest alternative branching patterns at some of the distal nodes, particularly in those cases in which few specimens are known, the major patterns of results and the similar results I have presented elsewhere (Good, 1987b) set the stage for meaningful discussions of such topics as history, taxonomy, and biogeography.

HISTORY OF GENERIC TAXONOMY IN THE GERRHONOTINAE

The evidence presented for previously published views on gerrhonotine taxonomy and relationships is presented below in the light of ancestral and derived character states in order to compare these views with the results of the present analysis. This chapter deals with phylogenetic hypotheses with respect to major groups of gerrhonotines; intrageneric hypotheses of relationships are discussed following the generic diagnoses in the Systematic Accounts below. As it is important to understand which species were known at the time each hypothesis was suggested, the reader should refer to the chronological list of species descriptions in Table 6. To avoid confusion, the species are referred to here by the genera used in the present study, not by those advocated by the worker under discussion.

The history of gerrhonotine systematics began with Wiegmann's (1828) description of the genus *Gerrhonotus*, containing the six species *Abronia deppii*, *A. taeniata*, *Elgaria coerulea*, *Barisia rudicollis*, *B. imbricata*, and *G. liocephalus*. In 1838, Gray divided this genus into four: *Abronia*, containing *A. deppii* and *A. taeniata*, defined as having the head depressed, a frontonasal present, and keeling reduced or absent; *Barisia*, with *B. rudicollis*, *B. imbricata*, and *B. lichenigera* (synonymous with *B. imbricata*), defined as lacking a depressed head and having prominent keeling and no frontonasal; *Elgaria*, including two species not known to Wiegmann, *E. kingii* and *E. multicarinata*, defined as lacking a depressed head and possessing a frontonasal, two pairs of "very narrow band-like anterior frontals" (the posterior internasals and supranasals), "scale-like" occipital scales, and some keeling; and *Gerrhonotus*, which retained *E. coerulea* and *G. liocephalus* and added *G. tesselatus* (synonymous with *G. liocephalus*) and *G. burnettii* (synonymous with *E. coerulea*), defined as having an undepressed head, a frontonasal present, and strong dorsal keeling. Gray retained these generic groupings in his 1845 catalogue, although he misspelled *Barisia* as *Barissia*.

With the exception of *Gerrhonotus*, all of these genera form monophyletic groupings as circumscribed by Gray. Two of the characters describing *Abronia*, the depressed head and reduced keeling, are derived features restricted to the genus. *Barisia* is diagnosed by the lack of a frontonasal and *Elgaria* by the condition of the posterior internasals and

TABLE 6. Chronological list of species descriptions

Year	Author	Species
1828	Wiegmann	*G. liocephalus*
		E. coerulea
		B. rudicollis
		B. imbricata
		A. deppii
		A. taeniata
1835	Blainville	*E. multicarinata*
1838	Gray	*E. kingii*
1864	Cope	*A. graminea*
1866	Cope	*M. antauges*
1869	Cope	*A. aurita*
1871	Bocourt	*M. moreleti*
		A. vasconcelosii
1873	Bocourt	*M. viridiflava*
1877	Peters	*C. rhombifer*
1878	Cope	*M. monticola*
1885	Günther	*A. oaxacae*
1890	Stejneger	*B. levicollis*
1913	Boulenger	*M. gadovii*
1934	Fitch	*E. cedrosensis*
		E. paucicarinata
1939	Martín del Campo	*A. ochoterenai*
1944	Tihen	*A. fuscolabialis*
1946	Hartweg and Tihen	*A. matudai*
1954	Tihen	*A. bogerti*
1958	Stebbins	*E. panamintina*
1961	Werler and Shannon	*A. reidi*
1963	Smith and Alvarez del Toro	*A. lythrochila*
1967	Bogert and Porter	*A. mixteca*
1970	McCoy	*G. lugoi*
1981	Smith and Smith	*A. chiszari*
1982	Campbell	*A. mitchelli*
1983	Hidalgo	*A. salvadorensis*
		A. montecristoi
1984	Campbell	*A. ornelasi*
1985	Good and Schwenk	*A. kalaina*
1985	Knight and Scudday	*E. parva*
1987	Karges and Wright	*M. juarezi*

Note: Species are referred to the genera used in the present study, not necessarily those originally used.

supranasals. *Gerrhonotus*, on the other hand, was defined by Gray solely on the basis of ancestral characters. Further, one of the two currently recognized species he placed in that genus, *E. coerulea*, possesses the derived characteristics of *Elgaria*.

Although he renamed most of them, Fitzinger (1843) recognized all of Gray's groupings except that he split *Abronia* into the two genera *Leiogerrhonotus* (containing *A. deppii*) and *Aspidosoma* (containing *A. taeniata*). No explanation was given.

Cope (1868), in the type description of *Abronia aurita*, contended that a regular gradation of characters connects Gray's *Abronia* with the rest of the gerrhonotines and therefore suggested that it should not be recognized. He did not discuss what those characters are.

O'Shaughnessy (1873), in a list of the species of gerrhonotines then known (Table 6), accepted only the single genus *Gerrhonotus*, but divided it into two groups. Group 1, was defined as having a depressed head, more-or-less swollen temporal and occipital elements, and reduced keeling, corresponded to Gray's *Abronia*. Group 2 was defined as lacking these characters. Group 1, as so defined, forms a monophyletic group with the derived features of Gray's *Abronia* and the further derived feature of swollen head scales. Group 2 was defined by ancestral characters. Group 2 was further subdivided on the basis of body proportions and the presence or absence of certain snout scales but O'Shaughnessy misaligned several species, undoubtedly through insufficient material available.

Cope (1878) reanalyzed gerrhonotine taxonomy and suggested that four genera should be recognized, based on the arrangement of snout scales. *Pterogasterus*, Gray's (1845) misspelling of Peale and Green's (1830) *Ptero-gastenes* (included by Gray in synonymy with *Gerrhonotus*), was accepted as including *G. liocephalus* and its synonyms as well as the newly described species *Mesaspis modesta*. It was defined by Cope by three pairs of internasal scutes (anterior and posterior internasals and supranasals) and the presence of a frontonasal and prefrontals. *Gerrhonotus*, containing Gray's (1838) *Elgaria, Abronia,* and the recently described species *Coloptychon rhombifer* and *M. monticola*, was defined as having two pairs of internasals and the frontonasal and prefrontals present. *Mesaspis* was newly erected for the species *M. moreleti* and *M. fulvus* (synonymous with *M. moreleti*). It was defined as having two pairs of internasals, a frontonasal present, and prefrontals lacking. *Barissia*, which corresponded to Gray's *Barisia/Barissia* with the addition of *M. antauges*, was defined as having two pairs of internasals, frontonasal lacking, and prefrontals present.

None of Cope's genera, with the exception of *Mesaspis*, which was monotypic, formed clearly monophyletic groups as he defined them. The only potentially derived feature in the definition of *Pterogasterus* is the presence of "three pairs of internasal scuta." This apparently refers to the medial expansion of the supranasals in addition to the primitively present anterior and posterior internasals. While this expansion is indeed derived and shared by his *Pterogasterus* species, it is also seen in his *Gerrhonotus* (*M. monticola, Coloptychon rhombifer,* and *Elgaria*), *Mesaspis*, and one of his *Barissia* (*M. antauges*). *Gerrhonotus*, as defined by Cope, possesses no derived features. Cope, however, misinterpreted scale homologies in this genus: in those species corresponding to Gray's *Elgaria*, his "two pairs of internasals" are the posterior internasals and the expanded supranasals (a derived condition). In those corresponding to Gray's *Abronia* and in *C.*

rhombifer, they are the anterior and posterior internasals (the ancestral condition). In *M. monticola*, they are the anterior internasals and the expanded supranasals (a derived condition). This latter condition is also seen in many specimens of Cope's *Mesaspis*. The other derived condition suggested in Cope's definition of *Mesaspis* was the lack of prefrontals. *Barissia* is apomorphic in the loss of the frontonasal. However, this condition in *M. antauges* is probably not homologous to that seen in the rest of Cope's *Barissia* and is shared with a variety of other species unknown to Cope. It is also by no means certain that it is even a consistent characteristic of *M. antauges*.

By 1900, Cope had restricted his view of generic diversity in the gerrhonotines to two genera: *Gerrhonotus*, defined, as in his 1878 work, by the presence of a frontonasal, and *Barissia*, which lacked this element. As so defined, *Abronia fimbriata*, a synonym of *A. aurita*, was included in *Barissia*, while all other *Abronia* were included in *Gerrhonotus*. This character is discussed above.

Dunn (1936) stated that "recent authors have disregarded *Mesaspis*, and I prefer to drop *Barissia*." He felt that these genera did not represent "natural groups" and that the characters used to diagnose them were unimportant.

The discussion to this point has covered works in which gerrhonotines were categorized without explicit reference to phylogeny. This is exemplified by Cope's (1878) statement that his *Gerrhonotus monticolus* (=*Mesaspis monticola*) and *Mesaspis fulvus* (=*M. moreleti*) "are probably nearly allied, but present a difference in the cephalic scutellation which is of generic value." Both of these genera contained other species so it is clear that Cope did not look on his genera as reflections of phylogenetic relationship (see also Cope, 1869). Although Fitch (1938) discussed relationships within *Elgaria*, the first worker to really emphasize the phylogenetic aspect in higher-level gerrhonotine systematics was Smith (1942).

In Smith's view, the single genus *Gerrhonotus* was divisible into five species groups. The "*deppii* group," containing those species here placed in *Abronia* as well as *Coloptychon rhombifer*, was characterized by a flattened head, poorly developed lateral fold, enlarged dorsals, and enlarged limb, neck, and head granules. The placement of *C. rhombifer* in this group was done with hesitation, as no specimens were examined. The "*antauges* group" consisted of the species here referred to *Mesaspis*. Smith defined it on the basis of a moderately well developed lateral fold, a single loreal, supranasals present and sometimes enlarged, head scales flat, complete superciliary series, and 45 or more transverse dorsals. *Gerrhonotus liocephalus* alone was placed in the "*liocephalus* group," which was characterized as having an elongate body and tail, at least two loreals, a complete superciliary series, and a frontonasal. The "*coeruleus* group," here referred to as *Elgaria*, was characterized by the fusion of the supranasals with the anterior internasals, contact of the rostral and the nasal, and presence of a frontonasal. The final group consisted of what is here called *Barisia*: the "*imbricata* group," defined on the basis of the loss of a frontonasal, convex head scales, and reduction of the superciliary series.

The characters Smith used to define his "*deppii* group" are all derived and diagnose a monophyletic group. His "*antauges* group" characters contain only two derived states: the partial reduction of the lateral fold (which is further reduced in *Abronia*) and the expansion of the supranasals sometimes seen in the group. This latter character, however, is shared with species scattered throughout his other species groups. All of the characters used to define the "*liocephalus* group" are ancestral. The fusion of the supranasals and anterior internasals described by Smith for the "*coeruleus* group" is instead a loss of the anterior internasals and concurrent expansion of the supranasals. This, as discussed above, is a derived feature. It is in turn responsible for the contact of the nasals with the rostral. The three characters defining the "*imbricata* group" are derived and diagnose a monophyletic group.

Smith was the first worker to propose an hypothesis of gerrhonotine relationship (Figure 19). He gave no explanation for why he favored this phylogeny.

Smith also pointed out that the frontonasal character of such importance to various previous workers (particularly Cope) is highly variable in some groups of gerrhonotines and he suggested that it is probably only useful in the latter three of the groups listed above.

In 1949, Tihen (1949a) again reanalyzed the gerrhonotine species (and incidentally introduced the name Gerrhonotinae to the literature). He concluded that these species should be divided into five genera: *Abronia, Barisia, Elgaria, Gerrhonotus*, and *Coloptychon*. His *Abronia, Elgaria, Gerrhonotus*, and *Coloptychon* corresponded to those genera as circumscribed in the present paper and each possesses derived character states. In his *Barisia*, however, he included the species here referred to both *Barisia* and *Mesaspis*. This *Barisia* (sensu lato) he defined as having a skull not widened or depressed, the frontal bone not in contact with the maxillae, pterygoid teeth absent or vestigial, dorsal osteoderms with a well defined, strongly thickened basal area, a moderately developed lateral fold, finely granular neck scales, 6-10 nuchal scales, anterior internasals present, a postrostral present or absent, a well differentiated subocular and postocular series, subocular-temporal contact, and 12-14 longitudinal ventral rows. Most of these characters are variable or are ancestral for the Gerrhonotinae. Those that are not (reduced pterygoid teeth, a somewhat reduced lateral fold, and differentiated suboculars and postoculars) are all shared with *Abronia*. The latter character is also shared with *Elgaria*.

The genus *Coloptychon* was newly erected by Tihen for the species *C. rhombifer*, which he considered to be the most primitive species of gerrhonotine (Figure 20). He also considered *Abronia* an early offshoot of the Gerrhonotinae because of the presence of a weak lateral fold, which he considered primitive (although it probably is not), and osteoderms reminiscent of the fossil genus *Peltosaurus*. However, a close relationship of *Peltosaurus* (a glyptosaurine) to the gerrhonotines is uncertain. Tihen made it clear that he considered *Abronia* to possess also several derived features so that it, unlike *Coloptychon*, does not exhibit a primitive general morphology. Of his three remaining genera, he stated that *Gerrhonotus* has more primitive characteristics than do *Elgaria* or *Barisia*, and he therefore considered the latter two to be sister taxa.

Also in 1949, Tihen (1949b) analyzed the species in his genus *Barisia*. He established three species groups: The "*moreleti* group," containing *Mesaspis viridiflava*, *M. moreleti*, and *M. monticola*, was defined as having an unpaired postmental, nasal bones in contact with each other, no pterygoid teeth, small size, a complete superciliary series, the upper postnasals present, a variable cantholoreal series, the frontonasal and postrostral present or absent, a moderately developed lateral fold, a dorsal stripe, and a speckled venter. The

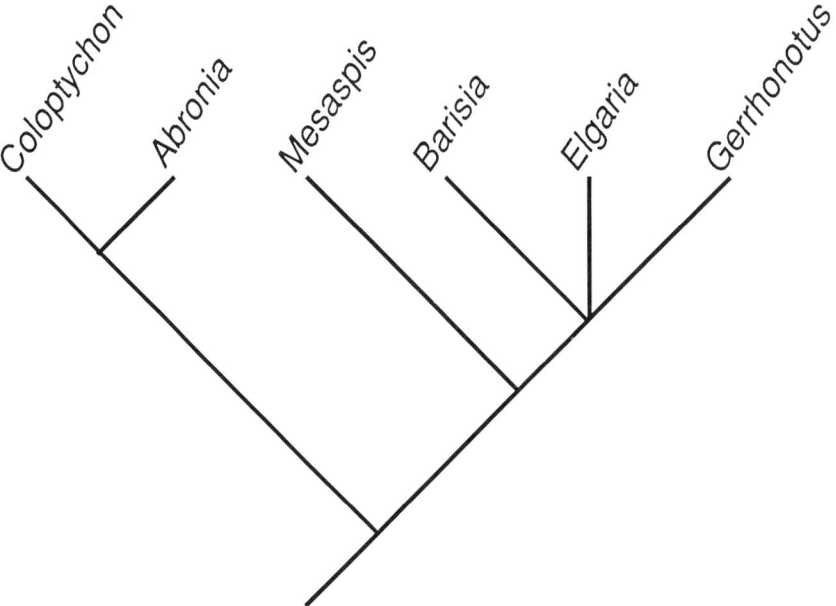

FIGURE 19. Gerrhonotine phylogeny as postulated by Smith (1942).

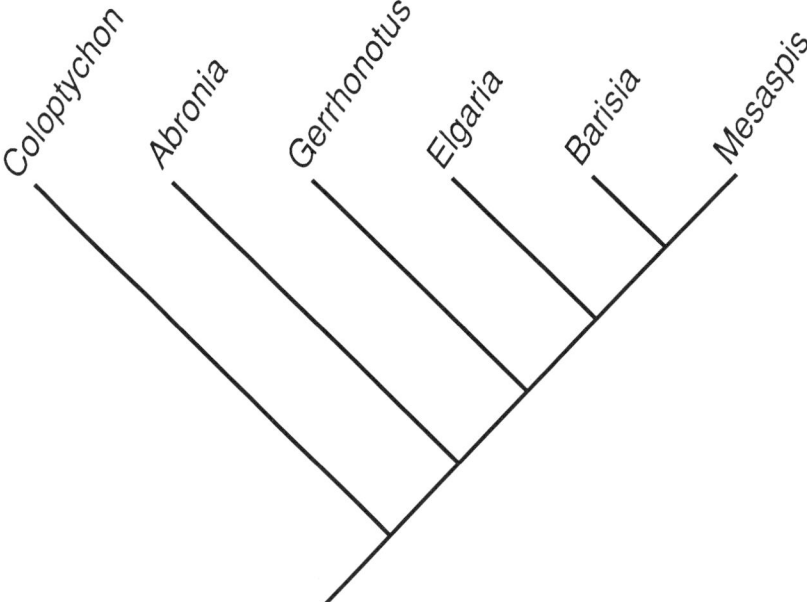

FIGURE 20. Gerrhonotine phylogeny as postulated by Tihen (1949a).

"*gadovii* group," containing *M. gadovii*, *M. antauges*, and *M. modesta* (=*M. antauges*), was defined as having paired postmentals, nasal bones separated from each other, small to moderate size, a dorsal stripe, a complete superciliary series, the upper postnasals present, two loreals (i.e., a loreal and a cantholoreal), anterior canthals present or not, the frontonasal and postrostral present or not, and a moderate lateral fold. The "*imbricata* group," with *B. imbricata*, *B. levicollis*, and *B. rudicollis*, had a paired postmental, separated nasal bones, vestigial pterygoid teeth, moderate to large size, a dorsal stripe, an unspotted venter, an incomplete superciliary series, no upper postnasals, a single loreal (=cantholoreal), the frontonasal and postrostral absent, and a well developed lateral fold.

Of the features defining Tihen's "*moreleti* group," unpaired postmentals, contact of the nasal bones, no pterygoid teeth, small size, spotted venter, and reduction of the lateral fold are derived. However, small size, spotted venter, reduced pterygoid teeth, and reduction of the lateral fold are shared with some or all of the "*gadovii* group" and/or *Abronia*. The "*gadovii* group," as defined by Tihen, has no derived features that are not also seen in the "*moreleti* group." The "*imbricata* group," however, is well defined as a natural group by the incomplete superciliary series, the condition of the cantholoreal series, and the lack of the upper postnasals and the frontonasal, all of which are derived features.

Tihen's views on the phylogeny of his *Barisia* are illustrated in Figure 21. In spite of the derived presence of an unpaired postmental, lack of pterygoid teeth, reduction of the prefrontals, and small size, he considered the "*moreleti* group" to be the most primitive in his genus *Barisia*. Tihen viewed the "*gadovii* group" as, in many ways, intermediate between the "*moreleti*" and "*imbricata*" groups.

Stebbins (1958), in his description of *Elgaria panamintina*, also examined the generic-level classification of the Gerrhonotinae. His view of phylogeny is illustrated in Figure 22. He accepted, although he did not explain why, Tihen's (1949a) *Abronia* and *Coloptychon*. He then combined Tihen's *Barisia* (sensu lato) and *Gerrhonotus* into the single genus *Gerrhonotus*, but erected within it two subgenera on the basis of general body form, habitat, and breeding biology: the subgenus *Gerrhonotus*, made up essentially of Tihen's *Gerrhonotus* and *Elgaria*, consisted of the more elongate, oviparous species inhabiting relatively warm, dry regions and the subgenus *Barisia* contained the stouter, ovoviviparous forms of relatively cool, moist areas. Thus defined, *E. coerulea* falls into the latter subgenus. Stebbins viewed the characters of scalation used by all previous workers as being too variable to be useful.

When defined in this way, the subgenus *Gerrhonotus* shows no derived features although habitat characteristics are hard to polarize. The subgenus *Barisia* is diagnosed by its body form and breeding biology, but both of these are seen also in *Abronia*. Placing *Elgaria coerulea* in the subgenus *Barisia* is warranted under the criteria set forth by Stebbins, but these criteria ignore much useful information which suggests that this does not reflect the phylogeny of the group.

Bogert and Porter (1967) agreed with Tihen's (1949a) view that *Abronia* is primitive.

Criley (1968) examined skulls from a wide variety of gerrhonotines in an attempt to discern phylogeny. He was unable to find any useful characters.

The view that Stebbins (1958) failed to take into account much useful information by excluding all scale characters was also held by Waddick and Smith (1974), who again

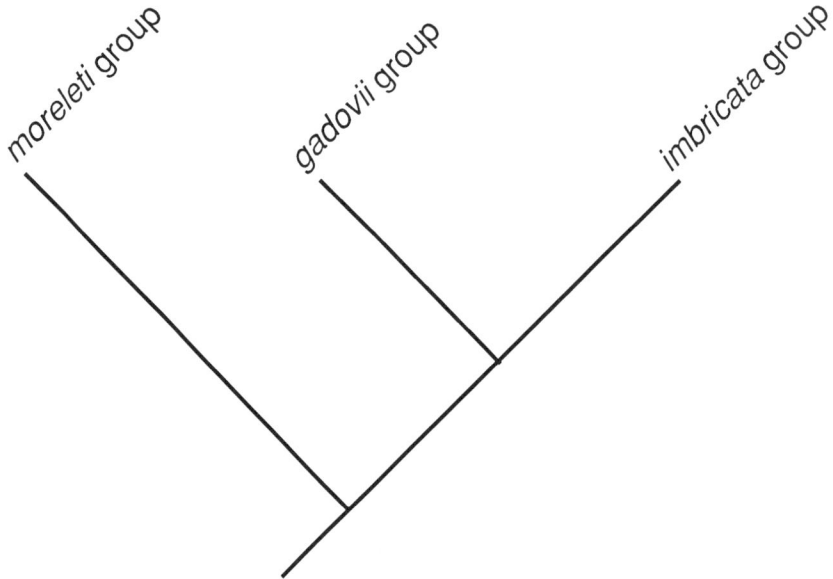

FIGURE 21. The phylogeny of *Barisia* (sensu Tihen, 1949a) as postulated by Tihen (1949b).

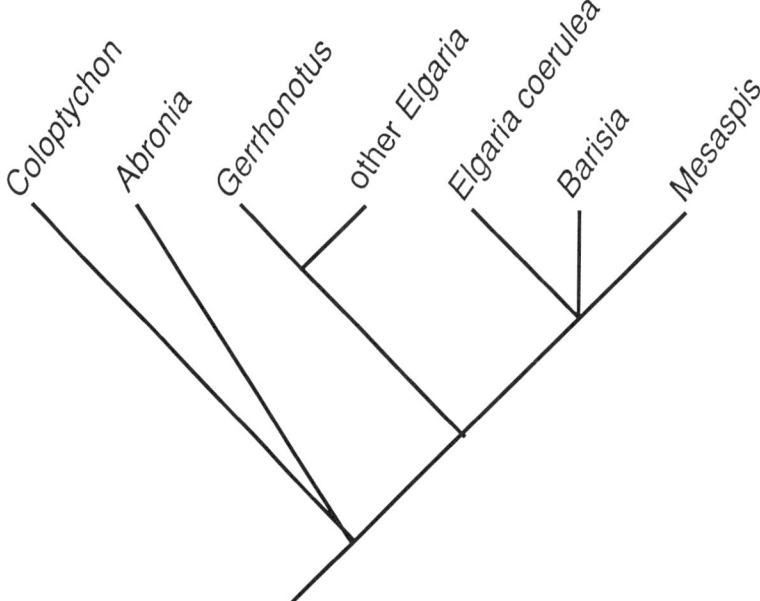

FIGURE 22. Gerrhonotine phylogeny as postulated by Stebbins (1958).

reanalyzed scale patterns in the Gerrhonotinae. However, they considered as a basic premise that all of Tihen's (1949a) genera were natural groups, for which there was no phylogenetic evidence. They also accepted Tihen's view that *Coloptychon* and *Abronia* are primitive offshoots of the Gerrhonotinae, and proceeded to analyze only Tihen's *Elgaria*, *Barisia*, and *Gerrhonotus* without further reference to them.

In Waddick and Smith's analysis, *Gerrhonotus* (sensu Tihen) was defined by the presence of a postrostral (and the resulting contact of the anterior internasals), absence of a cantholoreal, presence of anterior internasals (and the resulting nasal-rostral contact), lack of supranasal contact, and the presence of two anterior gulars. This last character is interspecifically variable; all others are ancestral. *Elgaria* was defined by the absence of a postrostral, presence of a cantholoreal, absence of anterior internasals, supranasal contact, and the presence of single anterior gulars. Of these characters, only the absence of the anterior internasals and the resulting (in the case of *Elgaria*) supranasal expansion are derived and restricted to the genus. The other derived features (absence of a postrostral and cantholoreal presence) are shared with their *Barisia* and *Abronia*. Aside from these two characters, their *Barisia* (which they admitted was the most "plastic" of the three genera) was defined as having the anterior internasals present, the supranasals not contacting, and two anterior gulars; all ancestral characters. The view held by Waddick and Smith that *Gerrhonotus* and *Elgaria* are sister taxa (Figure 23) is not upheld by their data.

Morafka (1977) agreed with Stebbins' (1958) view that *Barisia* (sensu lato) is insufficiently different from *Gerrhonotus* (including *Elgaria*) to warrant generic status. He suggested that *E. multicarinata* is almost intermediate between the two groups, but did not explain this statement. No discussion of *Abronia* or *Coloptychon* was provided.

Rieppel (1980), in his discussion of head musculature in the Anguimorpha, made the first attempt at an analysis of relationships among gerrhonotines on the basis of shared derived characters. He analyzed the gerrhonotine genera as defined by Tihen (1949a) and Waddick and Smith (1974) and thus, like these workers, assumed monophyletic groups without real evidence for them. On the basis of a few specimens each of *Barisia*, *Abronia*, *Elgaria*, and *Gerrhonotus*, he concluded that *Gerrhonotus* is the most primitive genus, with *Barisia* and *Elgaria* being the most closely allied of the remaining three (Figure 24). Despite the fact that no synapomorphies were described linking *Abronia*, *Barisia*, and *Elgaria*, *Gerrhonotus* was assigned ancestral status on the basis of Bogert and Porter's (1967) discussion and because of the presence of mostly ancestral character states in the genus. The only ancestral characters mentioned were the lack of a separate head on the *pseudotemporalis profundus* and the symmetrical position of the *levator pterygoidei* with respect to the epipterygoid. Both of these ancestral states were also seen in *Abronia*. On the basis of these characters, and because no consistent differences were observed between them, *Barisia* and *Elgaria* were considered to be sister taxa.

Rieppel was concerned mostly with higher-level systematics in his paper and thus used outgroups pertinent to all of the anguimorphs rather than to the gerrhonotines in particular. Since alternative gerrhonotine states are seen in the Diploglossinae and Anguinae, neither of the above characters is polarizable for the Gerrhonotinae and so yield no phylogenetically useful information.

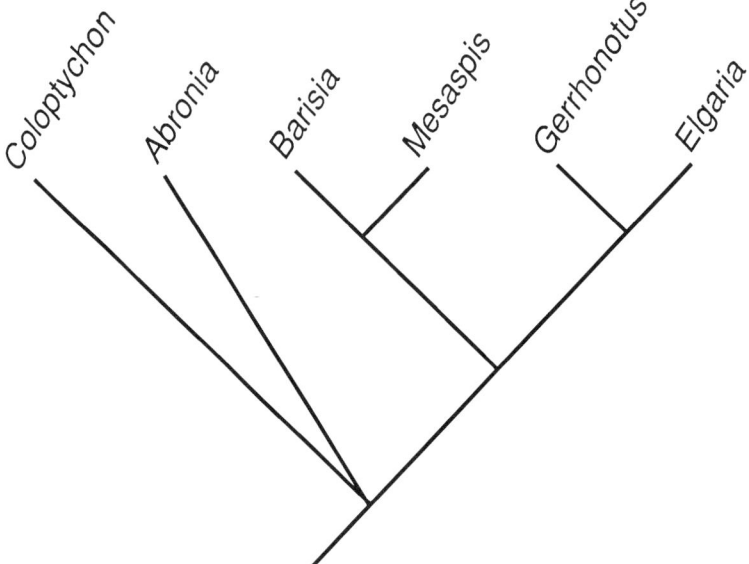

FIGURE 23. Gerrhonotine phylogeny as postulated by Waddick and Smith (1974).

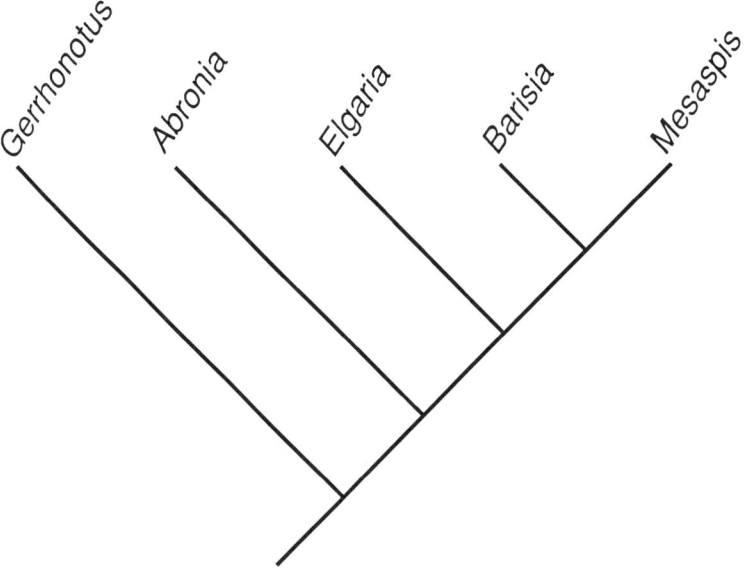

FIGURE 24. Gerrhonotine phylogeny as postulated by Rieppel (1980).

The most recent major work on the phylogeny of gerrhonotines was that of Gauthier (1982). He recognized that several of Waddick and Smith's (1974) characters were either ancestral or more widely inclusive than as described by them. He was also the first to suggest that *Abronia* may in fact be an offshoot of *Barisia* (sensu Tihen) rather than a primitive genus in the subfamily (Figure 25). In his view, the two genera are linked by the derived presence of pointed, delicate teeth that are recurved and widely spaced, contact between the anterior internasals, reduction or loss of pterygoid teeth, pronounced sexual dichromatism, and live birth. The condition of the teeth in *Barisia* is more variable than Gauthier recognized. The contact of the anterior internasals is likewise a poor synapomorphy since anterior internasal contact is the result of postrostral loss, a character which, along with cantholoreal presence, Gauthier used to link *Elgaria* with these genera. Anterior internasal contact is not seen in *Elgaria* because the anterior internasals are lacking. Sexual dichromatism and live-bearing remain good synapomorphies, although live-bearing is shared with *E. coerulea*.

Gauthier's diagnosis of *Abronia* consisted of five characters: fewer and larger dorsal scales, reduced body osteoderms, elongate limbs, a wide interpterygoid vacuity, and absence of pterygoid teeth. He suggested that many of these are juvenile characters and may be the result of the single, more inclusive character state of paedomorphosis.

Comparison of the results of the present analysis with some of the hypotheses of gerrhonotine relationships suggested by earlier workers is interesting. First and foremost, the hypothesis put forward by Smith (1942) and accepted by most subsequent workers that *Abronia* represents an offshoot of the early gerrhontine stock is not borne out. It appears to be, instead, as suggested by Gauthier (1982), more closely allied to Tihen's (1949a) *Barisia* (which includes *Mesaspis*). Tihen was correct, however, in his placement of *Coloptychon* outside of all other genera.

Smith (1942) viewed his "*antauges* group" (=*Mesaspis*) as being the most primitive gerrhonotine next to his "*deppii* group" (=*Abronia*). This also is not supported by the present analysis; *Mesaspis* and *Abronia* are here considered to be sister taxa and among the most highly derived forms in the subfamily.

Tihen (1949a) and Rieppel (1980) both saw *Gerrhonotus* as being outside of *Elgaria* and *Barisia* (sensu lato) while Waddick and Smith (1974) saw *Gerrhonotus* and *Elgaria* as sister taxa. This analysis supports the former view.

The division of *Gerrhonotus/Elgaria/Barisia/Mesaspis* into the two groups *Gerrhonotus/Elgaria* and *Barisia/Mesaspis*, as suggested by Stebbins (1958), is not supported. His inclusion of *E. coerulea* in the *Barisia/Mesaspis* clade on the basis of body form and breeding biology is also unwarranted.

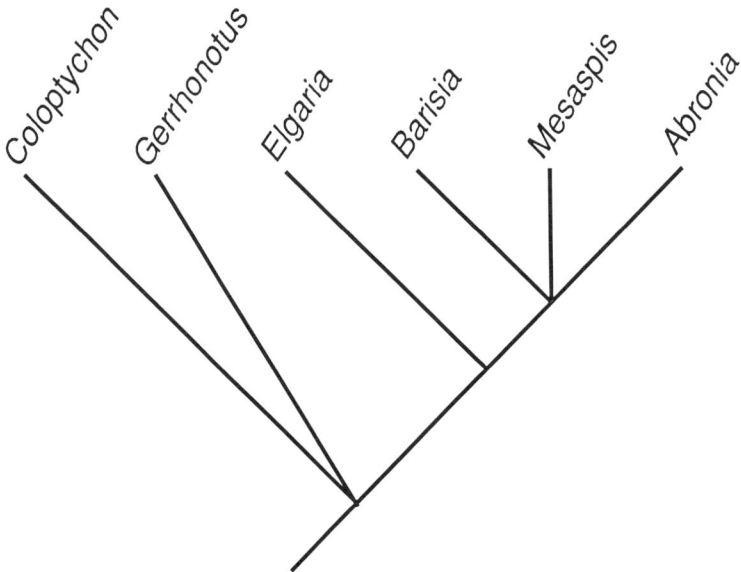

FIGURE 25. Gerrhonotine phylogeny as postulated by Gauthier (1982).

KEY TO THE SPECIES OF GERRHONOTINE LIZARDS

1. Arboreal lizards with relatively long, well-clawed limbs; lateral fold lacking between the forelimb and ear; vertical temporal scale rows 4 or fewer; transverse dorsals fewer than 40; nuchals 6 or fewer (except in *Abronia bogerti* and *A. chiszari*, each with 8) 2

1. Primarily terrestrial lizards with relatively short limbs; lateral fold between forelimb and ear present; vertical temporal scale rows 5; transverse dorsals more than 40 (except in *Barisia imbricata* and *B. rudicollis*); nuchals 8 or more (except in *B. rudicollis*, with 4-6) 20

2. Postmental 1; suboculars usually 3; contact of penultimate supralabial with the orbit through loss of the posterior supralabial usually present; high elevation species east of the Isthmus of Tehuantepec 3

2. Postmentals 2; suboculars usually 2; posteriormost supralabial usually not lost; species west of the Isthmus of Tehuantepec 9

3. Protuberent supra-auriculars; pre-auriculars granular 4

3. Supra-auriculars not protuberent; pre-auriculars not granular 8

4. Lateral longitudinal ventrals expanded to almost twice the width of the other ventrals 5

4. All longitudinal ventral rows more-or-less equal in width 6

5. Longitudinal dorsal scale rows 16; primary temporals 3; Volcán Tacaná and Volcán Tajumulco, Mexico and Guatemala .... *Abronia matudai*

5. Longitudinal dorsal scale rows 14; primary temporals 4-5; Sierra Madre de Chiapas, Mexico .... *Abronia ochoterenai*

6. Posterior infralabial expanded to almost twice the length of the other infralabials 7

6. Posterior infralabial more-or-less equal to the other infralabial ....*Abronia aurita*

7. Anterior canthal usually (?) lacking; labials blood red; Chiapas, Mexico .... *Abronia lythrochila*

7. Anterior canthal present; labials not blood red; south-central Guatemala .... *Abronia vasconcelosii*

8. Interoccipitals more than 1; longitudinal ventrals 12; superciliary- cantholoreal contact present; 3 large and 1 small pair of chinshields .... *Abronia montecristoi*

8. Interoccipital 1; longitudinal ventrals 14; supercilary-cantholoreal contact lacking; 4 pairs of large chinshields .... *Abronia salvadorensis*

9. Nasal often contacting the third supralabial; posterior internasals approximately twice the size of the anterior; anterior canthal lacking; postoccipitals rugose; dorsal crossbands 6-8 10

9. Nasal-third supralabial contact lacking; anterior and posterior internasals approximately equal in size; anterior canthal present; postoccipitals relatively smooth; dorsal crossbands 10-11 14

10. Strongly knoblike posterior head scales; 4 pairs of large chinshields; longitudinal dorsals 10-13; longitudinal dorsals at hind limbs 6; lateral dorsals posteriomedially rounded and in rows oblique to the lateral fold; osteoderms reduced or lacking on the dorsum; lateral fold reduced; scales on the trailing edges of the limbs not granular or subgranular 11

10. Posterior head scales not strongly knoblike; 3 pairs of large and 1 pair of small chinshields; longitudinal dorsals 14; longitudinal dorsals at hind limb 8; lateral dorsals not posteromedially rounded, in rows perpendicular to the lateral fold; osteoderms present on the dorsum (in adults); lateral fold stronger; scales on trailing edge of limbs granular or subgranular 13

11. Fourth temporal row often lost; third primary temporal often lost; interoccipitals 3; postmentals reduced in size; suboculars 2; lowermost primary temporal not fused with antepenultimate supralabial 12

11. Fourth temporal row present; third primary temporal present; interoccipital 1; postmentals not reduced; subocular 1; lowermost primary temporal fused with the antepenultimate supralabial .... *Abronia deppii*

12. Frontonasal large; tertiary temporals 4; antepenultimate supralabial usually contacting the orbit; nuchals 6 .... *Abronia mixteca*

12. Frontonasal small or absent; tertiary temporals 2; penultimate supralabial contacting the orbit; nuchals 4 .... *Abronia oaxacae*

13. Color bright green in the male, duller green in the female; crossbands faint or lacking in the adult; nuchals often 4; Veracruz and Puebla, Mexico .... *Abronia graminea*

13. Color white or off-white with 6-8 very distinct bluish to black dorsal crossbands; Sierra Madre Oriental from Hidalgo to Tamaulipas, Mexico .... *Abronia taeniata*

14. Subocular-temporal contact present; supranasals at least somewhat expanded; fourth temporal row unreduced; secondary temporals 4 15

14. Subocular-temporal contact lacking; supranasals unexpanded; fourth temporal row reduced; secondary temporals 3 17

15. Midline contact of the supranasals; 3 primary temporals contacting the orbit; interoccipital 1; postoccipital rows 2 16

15. Midline contact of the supranasals lacking; 2 primary temporals contacting the orbit; interoccipitals 2; postoccipital row 1 .... *Abronia mitchelli*

16. Prefrontal-superciliary contact present; osteoderms reduced on the dorsum; anterior canthal lacking .... *Abronia reidi*
16. Prefrontal-superciliary contact lacking; osteoderms present; anterior canthal present .... *Abronia ornelasi*
17. Anterior superciliary almost twice the length of the other superciliaries; lower primary temporals expanded so that only 2-3 primary temporals are present; nuchals 8; longtitudinal ventral rows 12 18
17. Anterior superciliary more-or-less the length of the other superciliaries; lower primary temporals not expanded; primary temporals 4; longitudinal ventral rows 14 19
18. Anterior internasals may be longitudinally divided; primary temporals 2; southwestern Oaxacan highlands, Mexico .... *Abronia bogerti*
18. Anterior internasals not divided; primary temporals 3; supposedly Sierra de los Tuxtlas, Veracruz, Mexico .... *Abronia chiszari*
19. Midline contact of the frontoparietals present; partial fusion of the frontoparietals and frontal; midline contact of the second pair of chinshields; dorsal osteoderms reduced; bright turquoise color; Cerro Pelón, Oaxaca, Mexico .... *Abronia kalaina*
19. Frontoparietal contact absent; frontoparietal-frontal fusion absent; chinshield contact absent; dorsal osteoderms not reduced; green color; Cerro Zempoaltepec, Oaxaca, Mexico .... *Abronia fuscolabialis*
20. Anterior internasals absent 21
20. Anterior internasals present 27
21. Longitudinal dorsals 16; venter immaculate or with longitudinal stripes between the scale rows 22
21. Longitudinal dorsals 14; venter with longitudinal stripes or speckles in the middle of the scale rows 23
22. Body size very small; keels lacking; infralabials 11-12; suffuse dark pigmentation between dorsal crossbands; Nuevo León, Mexico .... *Elgaria parva*
22. Body size moderate to fairly large; keeling strong or reduced, but always present; infralabials 8-10; dorsal crossbands lacking or, if present, without suffuse pigmentation between; western United States and Canada .... *Elgaria coerulea*
23. Lower subocular somewhat triangular; dorsal crossbands extending into the lateral fold; keeling weak, lacking on the limbs 24
23. Lower subocular not triangular; dorsal crossbands not extending into the lateral fold; keeling strong, present on the limbs .... *Elgaria multicarinata*
24. Suffuse dark pigmentation between prominent dorsal crossbands; United States and mainland Mexico 25
24. Dorsal crossbands usually weak and always lacking suffuse pigmentation; Baja California 26
25. Body elongate, limbs relatively short; labials with distinct black-and-white markings; venter spotted; New Mexico, Arizona, and western Mexico .... *Elgaria kingii*
25. Body not elongate, limbs not short; labials without distinct markings; venter with longitudinal stripes; southeastern California .... *Elgaria panamintina*

26. Crossbands often indistinct; ventral stripes distinct; Cedros Island .... *Elgaria cedrosensis*

26. Crossbands usually relatively distinct; ventral stripes often indistinct; mainland Baja California .... *Elgaria paucicarinata*

27. Suboculars differentiated from pre- and postoculars; 3 large and 1 small pair of chinshields present; sublabials 4-5; caudal whorls fewer than 100; scales per whorl (near base of tail) 15-17; species usually sexually dichromatic; high-elevation species 28

27. Suboculars undifferentiated from pre- and postoculars; 4-5 pairs of large chinshields; sublabials 5-7; caudal whorls more than 100; scales per whorl 20-24; not sexually dichromatic; low- to moderate-elevation species 36

28. Supranasal and upper postnasal fused; frontonasal lacking; superciliaries 3 or fewer; snout short and deep; lateral fold strong 29

28. Supranasal and upper postnasal not fused; frontonasal usually present; superciliaries 5 or more; snout not short and deep; lateral fold somewhat reduced 31

29. Nasal-rostral contact present; keeling very strong, scales acuminate; postoccipitals keeled; nuchals 4-6; longitudinal ventrals 14; limbs relatively long and well clawed; possibly arboreal .... *Barisia rudicollis*

29. Nasal-rostral contact lacking; keeling present but scales not acuminate; postoccipitals not keeled; nuchals 8; longitudinal ventrals 12; limbs shorter; terrestrial 30

30. Superciliary 1; longitudinal dorsals 16; transverse dorsals more than 40 .... *Barisia levicollis*

30. Superciliaries 3; longitudinal dorsals 14; transverse dorsals fewer than 40........... .... *Barisia imbricata*

31. Postmental 1; lateral supraoculars often 2 32

31. Postmentals 2; lateral supraoculars 3 34

32. Longitudinal dorsals 18-20; nuchals 10 .... *Mesaspis moreleti*

32. Longitudinal dorsals 14-16; nuchals 8 33

33. Longitudinal dorsals 14; frontonasal usually absent; anterior internasals often longitudinally divided; canthal-cantholoreal fusion .... *Mesaspis viridiflava*

33. Longitudinal dorsals 16; frontonasal present; anterior internasals not longitudinally divided; canthal and cantholoreal separate .... *Mesaspis monticola*

34. Body size small; keeling reduced or absent; supranasal somewhat expanded; suboculars 2; postrostral present; frontal-parietal contact broad; anterior superciliary almost twice the length of the other superciliaries 35

34. Body size moderate; keeling present; supranasal unexpanded; subocular 1; postrostral absent; frontal-parietal contact narrow; superciliaries more- or-less equal in length .... *Mesaspis gadovii*

35. Posterior internasal divided; cantholoreal sometimes partially fused to preocular; male much more robust than female; Pico de Orizaba, Veracruz .... *Mesaspis antauges*

35. Posterior internasal not divided; cantholoreal and preocular not fused; male similar to female in robustness; Sierra de Juárez, Oaxaca .... *Mesaspis juarezi*

36. Postrostrals probably usually 2; 5 pairs of large chinshields; scales per vertical temporal row 5; nuchals 12; keeling lacking; longitudinal ventrals 10; suboculars more than 3 .... *Coloptychon rhombifer*

36. Postrostrals 0-1; 4 pairs of large chinshields; scales per vertical temporal row 4; nuchals 10; keeling present; longitudinal ventrals 12-14; suboculars 3 or fewer 37

37. Body size small; keeling reduced; longitudinal ventrals 14; postrostral absent; suboculars 3 .... *Gerrhonotus lugoi*

37. Body size large; keeling strong; longitudinal ventrals 12; postrostral present; suboculars 2 .... *Gerrhonotus liocephalus*

SYSTEMATIC ACCOUNTS

Coloptychon Tihen

Gerrhonotus, Peters, 1877, Monatsb. Akad. Wiss. Berlin, 1876: 299. Part.
Coloptychon Tihen, 1949, Am. Midl. Nat. 41: 584.

Type species. Coloptychon rhombifer (Peters).
Derived features. The following derived gerrhonotine scale features diagnose *Coloptychon*: two longitudinally arranged postrostrals, expanded supranasals (probably usually to the midline), prefrontal-superciliary contact, five pairs of large chinshields, a reduced lateral fold, and reduced dorsal osteoderm development. In addition, three features whose polarity is unclear may also diagnose the genus: contact of the nasal with the third supralabial, a high number of suboculars (3-5 rather than two), and an increase from 9-12 to 13-14 supralabials.
  Of the above characters, only the presence of two postrostrals and five pairs of large chinshields are restricted to *Coloptychon*; all others are shared through homoplasy with other gerrhonotines. On the other hand, *Coloptychon* lacks a number of derived features seen in all other genera (see below): a reduction in canthal/loreal number, a reduction in temporal number, a reduction in nuchals, an increase in keeling, and an increase in the number of longitudinal ventrals. Reduction in subocular number, of questionable polarity, may be another such character.
  Certain derived aspects of coloration are also diagnostic of *Coloptychon*. These are primarily the characteristic dorsal and ventral crossbanding patterns described below.

Coloptychon rhombifer (Peters)

Gerrhonotus rhombifer Peters, 1877, Monatsb. Akad. Wiss. Berlin, 1876: 299.
Coloptychon rhombifer, Tihen, 1949, Am. Midl. Nat. 41: 585.

Holotype. Zoologisches Museum, Berlin, 8655; Chiriquí, Panama.
Derived features. As for the genus, above.

Color pattern. The color of *C. rhombifer* in life is unknown to me; in alcohol, tans and browns predominate. The dorsum is patterned with eight triangular or rhomboidal light crossbands on a darker background. The ventral pattern consists of approximately six narrow, posteriorly pointed chevrons on the anterior trunk, throat, and chin. The posterior venter is darker and unbanded. The temporal region is of the same dark coloration as the ground color of the dorsum; the top of the head and labials are lighter. The labials also show two dark bands extending from the chevrons of the chin. The tail is similar in pattern to the dorsum. The limbs are mottled.

Distribution. *Coloptychon rhombifer* is known from only three specimens, one from western Panama and the others from eastern Costa Rica (Figure 26). The species apparently occurs at low elevations, but nothing is known of its habits or habitat.

Discussion. Tihen (1949a) erected the genus *Coloptychon* for this species, without having seen any specimens, on the basis of the descriptions by Peters (1877) and Bocourt (1878) and of Bocourt's illustration of head scutellation. However, some of the characteristics cited by these authors, such as enlarged neck scales and lack of prefrontal-superciliary contact, do not occur in *Coloptychon*. Many of the characters actually occurring in the genus are ancestral among gerrhonotines, and it is therefore largely identified by the lack of the synapomorphies seen in the rest of the subfamily.

Gerrhonotus Wiegmann

Gerrhonotus Wiegmann, 1828, Isis, 21: 379. Part.
Scincus, Peale and Green, 1830, J. Acad. Nat. Sci. Phila. 6: 233. Part.
Ptero-gastenes Peale and Green, 1830, ibid., p. 234.
Pterogasterus Gray, 1845, Cat. Spec. Liz. Coll. British Mus., p. 53.

Type species. *Gerrhonotus tessellatus* Wiegmann = *G. liocephalus* Wiegmann.

Derived features. Derived gerrhonotine features that are shared among *Gerrhonotus* species are as follows: loss of at least one canthal/loreal, reduction in the number of temporals per vertical row from 5-6 to four or fewer, acquisition of keeling, reduction in the number of nuchals from 12 to ten or fewer, four large and one small pair of chinshields, an increase from ten to 12-14 longitudinal ventrals, and a characteristic light-on-dark dorsal crossbanding pattern. Of these, all but the characteristic chinshield arrangement and color pattern are shared with all other gerrhonotine genera except *Coloptychon*; these two, however, are restricted to *Gerrhonotus* and diagnose the genus. A further diagnostic feature shared only with *Mesaspis moreleti* is an increase in longitudinal dorsals to 18-20.

Two other characters of questionable polarity may also diagnose *Gerrhonotus*, although neither is restricted to the genus. These are the contact of the nasal and the third supralabial seen also in *Coloptychon* and some *Abronia*, and an increase from 9-12 to 13-14 supralabials seen also in *Coloptychon*.

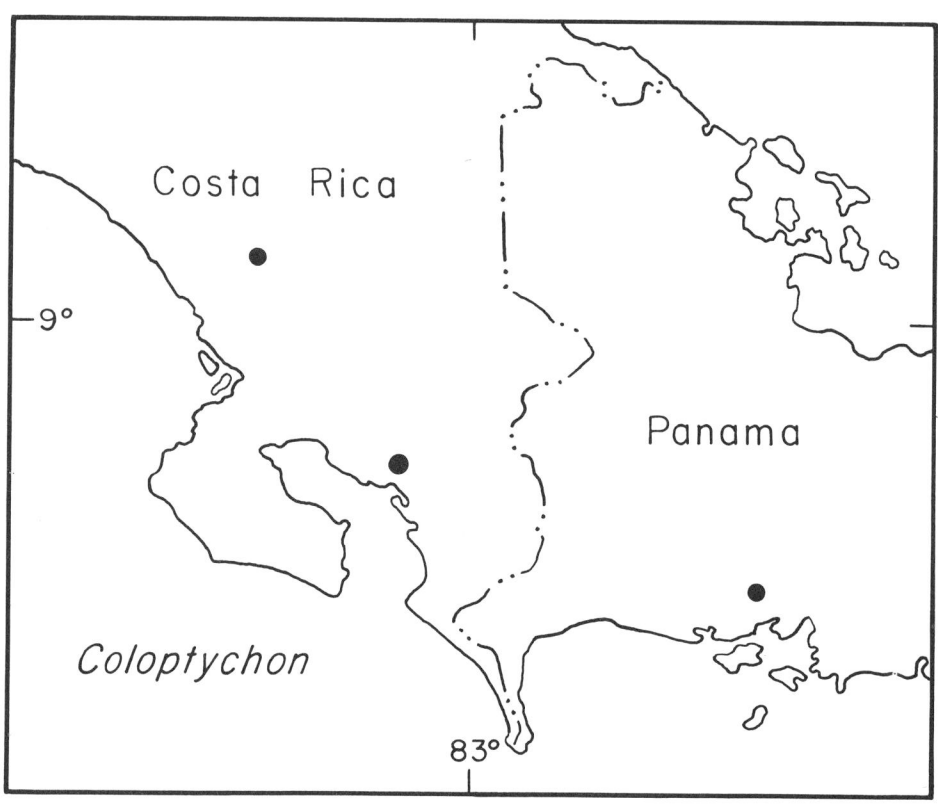

FIGURE 26. The geographic distribution of *Coloptychon*.

Gerrhonotus liocephalus Wiegmann

Gerrhonotus liocephalus Wiegmann, 1828, Isis, 21: 381.
Scincus ventralis Peale and Green, 1830, J. Acad. Nat. Sci. Phila. 6: 233.
Ptero-gastenes ventralis Peale and Green, 1830, ibid., p. 234.
Gerrhonotus tessellatus Wiegmann, 1834, Herp. Mex., pt. 1, p. 32.
Gerrhonotus infernalis Baird, 1858, Proc. Acad. Nat. Sci. Phila. 10: 255.
Gerrhonotus ophiurus Cope, 1866, Proc. Acad. Nat. Sci. Phila. 18: 321.
Gerrhonotus lemniscatus Bocourt, 1871, Bull. Nouv. Arch. Mus. 7:105.
Pterogasterus infernalis, Cope, 1878, Proc. Amer. Philos. Soc. 17: 96.
Pterogasterus ophiurus, Cope, 1878, ibid.
Pterogasterus tessellatus, Cope, 1878, ibid.
Pterogasterus ventralis, Cope, 1878, ibid.
Gerrhonotus leiocephalus, Dunn, 1936, Proc. Acad. Nat. Sci. Phila. 88: 475.

Holotype. Zoologisches Museum, Berlin, 1153; Mexico; restricted to Tlapancingo, Oaxaca, by Smith and Taylor (1950).

Derived features. Of the numerous characters in which *G. liocephalus* and the only other member of the genus *Gerrhonotus*, *G. lugoi*, differ, none are derived and shared by all *G. liocephalus*. There are a few characters, however, that show derived states in some *G. liocephalus*: supranasal expansion, prefrontal-superciliary contact, a single preocular, loss of one canthal/loreal element, and two temporals contacting the orbit. In addition, two characters of uncertain polarity may aid in diagnosing the species: postrostral presence and presence of two suboculars rather than three.

Color pattern. The adult color pattern of *G. liocephalus* is a tan or yellowish- to reddish- brown background with 8-10 lighter, narrow crossbands on the dorsum. The venter and head are unmarked. The tail is similar in pattern to the dorsum.

Distribution. *Gerrhonotus liocephalus* occurs from central Texas south through eastern and southern Mexico as far as Chiapas (Figure 27). It is a species of varied habitat, being found over a large elevational range from semidesert regions to pine forest. The subspecies *ophiurus* inhabits lowland tropical forest.

Discussion. Six subspecies (*austrinus*, *infernalis*, *liocephalus*, *loweryi*, *ophiurus*, and *taylori*) have been described. Although, for the purposes of this analysis, all of these are considered conspecific, it is likely that *G. liocephalus* as currently recognized is comprised of several species, since much geographic variation in morphology and habitat occurs. The most likely candidates for specific status are *infernalis*, *liocephalus*, and *ophiurus*, which differ from each other in a suite of characters. Further work is necessary to shed light on this problem.

Gerrhonotus lugoi McCoy

Gerrhonotus lugoi McCoy, 1970, Southwest. Nat. 15: 37.
Barisia lugoi, Waddick and Smith, 1974, Great Basin Nat. 34: 264.

Holotype. Carnegie Museum, Pittsburgh, 49012; northern tip Sierra de San Marcos, approximately 11 km SW Cuatro Ciénegas de Carranza, Coahuila, Mexico, ca. 800 m.

Derived features. By far the most obvious diagnostic feature of *G. lugoi* is its small size; it can in many respects be thought of as a miniaturized *G. liocephalus*. Other derived scale characters which differentiate *G. lugoi* from *G. liocephalus* are 14 longitudinal ventral rows and loss of keeling. The characters of uncertain polarity which may aid in diagnosing *G. lugoi* are listed in the diagnosis of *G. liocephalus* above.

Color pattern. The color pattern of *G. lugoi* is essentially as described for *G. liocephalus*, above.

Distribution. *Gerrhonotus lugoi* is known from only three specimens, all collected in the vicinity of the type locality near Cuatro Ciénegas, Coahuila, Mexico (Figure 27). The species apparently inhabits desert canyons at lower elevations than *G. liocephalus* in the same general region.

Discussion. Although correctly placed in *Gerrhonotus* by McCoy (1970), this species was listed as a species of *Barisia* by Waddick and Smith (1974) and Smith (1986), their concept of *Barisia* including both the genera *Barisia* and *Mesaspis* as here

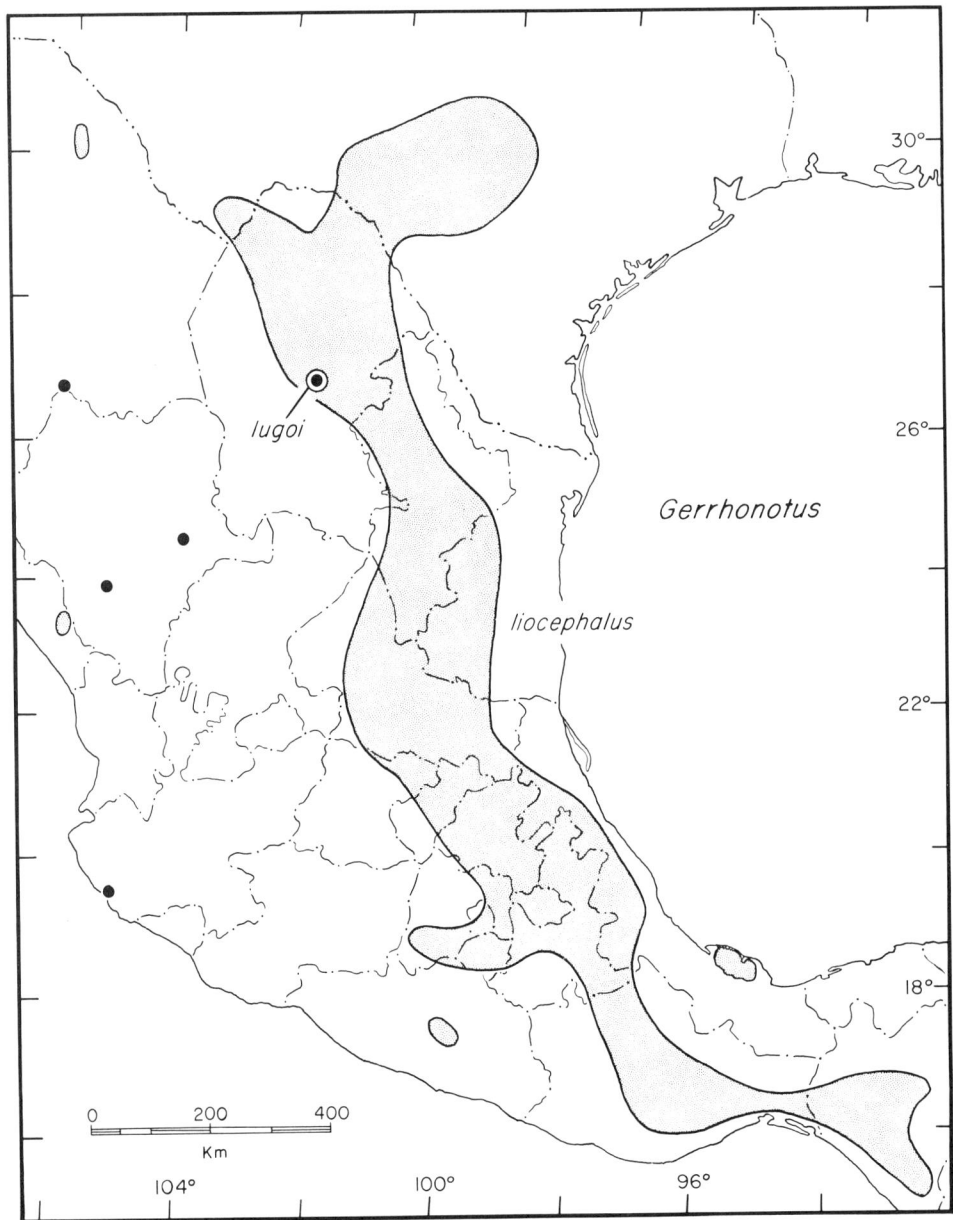

FIGURE 27. The geographic distribution of *Gerrhonotus*.

circumscribed. All of the characters discussed by Smith (1986) as allying *G. lugoi* with *Barisia* and *Mesaspis* are ancestral for the Gerrhonotinae.

Morafka (1977) suggested that *G. lugoi* might represent simply the end of an altitudinal gradient in various characters within *G. liocephalus*.

Elgaria Gray

Gerrhonotus Wiegmann, 1828, Isis, 21: 379. Part.
Cordylus, Blainville, 1835, Nouv. Ann. Mus. Nat. Hist. Natur. 4: 57. Part.
Elgaria Gray, 1838, Ann. Mag. Nat. Hist., ser. 1, 1: 390.
Trachypeltis Fitzinger, 1843, Syst. Reptil., p. 21.
Tropidolepis, Skilton, 1849, Amer. J. Sci. Arts 7: 202. Part.

Type species. Elgaria multicarinata (Blainville).

Derived features. A number of derived gerrhonotine scale features are characteristic of *Elgaria*, most of which are shared with other genera. Those shared with all other gerrhonotine genera with the exception of *Coloptychon* are listed in the diagnosis of *Gerrhonotus* above. Derived gerrhonotine features of *Elgaria* seen also in *Barisia*, *Mesaspis*, and *Abronia* are the loss of all postrostrals, fusion of the posterior canthal and the posterior loreal to form a single cantholoreal (seen also in some *G. liocephalus*), presence of a single preocular (seen also in some *G. liocephalus*), reduction from 11-12 to 8-10 infralabials and from 7-8 to 4-5 sublabials, and granular scales on the trailing edges of the limbs.

A single derived scale character is unique to *Elgaria*: the loss of the anterior internasals. In addition, the characteristic dorsal crossbanding, longitudinal ventral striping, and head spotting seen in the genus are diagnostic.

Discussion. Elgaria species are quite invariant in scale characters relative to intrageneric variation in other alligator lizard genera. Fitch (1938), in his analysis of variation in the genus, relied primarily on differences in mean scale counts and on color pattern. Considerable overlap among species occurs in most of these characters, and the species are often best distinguished by distribution and habitat.

Fitch's (1938) views on relationships within *Elgaria* were never explicitly stated, although he implied in his discussion what he thought these relationships were. He suggested that *E. coerulea* was the earliest offshoot (making up his "*coeruleus* group"), with *E. multicarinata, E. paucicarinata, E. cedrosensis,* and *E. kingii* (his "*multicarinatus* group") being linked by the presence of a similar crossbanded color pattern, an undivided occipital, an unreduced frontonasal, a long tail, a pale iris, and oviparity, as well as similarities in habitat and foraging habits. He next linked *E. cedrosensis, E. paucicarinata,* and *E. kingii* on the basis of an increased number of dorsal and ventral scale rows, reduction in keeling, characteristic spots on the labials, and characteristic black and white coloration of the lateral fold. *Elgaria cedrosensis* and *E. paucicarinata* were considered closely related because of their general similarity; no specific characters were discussed. *Elgaria panamintina* and *E. parva* were unknown to Fitch.

Of Fitch's characters linking the "*multicarinatus* group," only color pattern and possibly eye color are derived. The head scale characters mentioned are variable intraspecifically, and therefore of little value. Such characters as habitat and foraging habits are difficult to polarize among gerrhonotines because of variability in the outgroups.

While high numbers of transverse dorsal and ventral scale rows are ancestral for the Gerrhonotinae, it is likely that the condition in *E. kingii*, *E. cedrosensis*, and *E. paucicarinata* is derived within *Elgaria*. Fitch's other three characters linking these species are also likely derived.

Smith and Taylor (1950) divided *Elgaria* (or at least those forms occurring in Mexico; i.e., excluding *E. coerulea*) into two groups, the "*kingii* group," possessing black and white labial markings, and the "*multicarinata* group," lacking such markings. As so defined, the "*multicarinata* group" lacks phylogenetic diagnosis.

Banta (1963) suggested that *E. multicarinata*, *E. paucicarinata*, *E. cedrosensis*, and *E. panamintina* are more closely related to each other than any is to *E. kingii*. No explanation was given for this belief.

I have elsewhere (Good, 1988a) analyzed allozyme variation among all species of *Elgaria* except *E. parva* and *E. cedrosensis*, and I agree with Fitch's (1938) hypothesis of relationships among the species known to him. In addition, I have hypothesized a close relationship of *E. panamintina* to *E. kingii* relative to the other species analyzed.

Elgaria cedrosensis (Fitch)

Gerrhonotus cedrosensis Fitch, 1934, Copeia 1934: 6.
Elgaria cedrosensis, Tihen, 1949, Amer. Midl. Nat. 41: 595.

Holotype. California Academy of Sciences, San Francisco, 56187; canyon on southeast side of Cedros Island, Baja California Sur, Mexico.

Derived features. *Elgaria cedrosensis* possesses the following derived *Elgaria* features: a somewhat triangular lower postocular (seen also in *E. kingii*, *E. panamintina*, and *E. paucicarinata*), 14 longitudinal ventral scale rows (seen in all *Elgaria* except *E. coerulea*), a reduction in keeling (seen also in *E. kingii*, *E. panamintina*, *E. parva*, and *E. paucicarinata*) and black-and-white markings in the lateral fold (also in *E. kingii*, *E. panamintina*, *E. parva*, and *E. paucicarinata*) and on the supralabials (in *E. kingii* and *E. paucicarinata*, and weakly in *E. parva*). Additional characters for which the polarity is unclear are a relatively long, slender body form and ventral stripes running along the scale rows. The former character is shared by all *Elgaria* except *E. coerulea* and the latter by all except *E. coerulea* and *E. parva*. No derived features that are restricted to *E. cedrosensis* have been observed.

Color pattern. The dorsal color pattern of *E. cedrosensis* is characterized by indistinct to almost obsolete dark crossbands on a background of gray-brown. The sides are distinctly banded with black. The head is almost immaculate or lightly spotted; the labials are characteristically patterned with black-and-white. The venter is lighter than the dorsum and has dark stripes running along the scale rows. The tail and limbs are virtually patternless.

Distribution. *Elgaria cedrosensis* is known only from Cedros Island, Baja California Sur, Mexico (Figure 28).

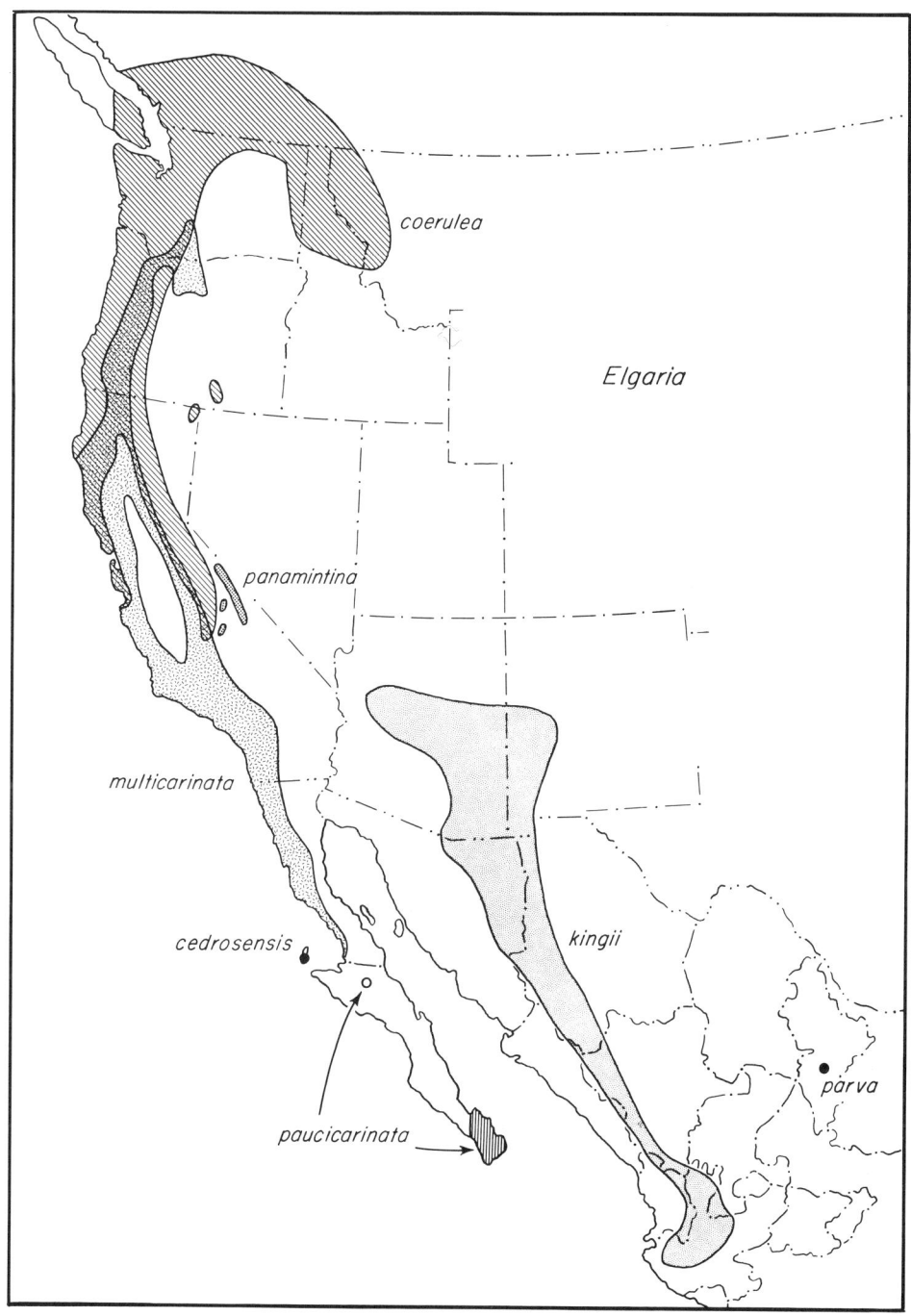

FIGURE 28. The geographic distribution of *Elgaria*.

Elgaria coerulea (Wiegmann)

Gerrhonotus coeruleus Wiegmann, 1828, Isis, 21: 380.
Gerrhonotus burnettii Gray, 1831, Cuv. An. King., Reptilia, p. 64.
Gerrhonotus caeruleus Gray, 1845, Cat. Spec. Liz. Coll. British Mus., p. 54.
Elgaria formosa Baird and Girard, 1852, Proc. Acad. Nat. Sci. Phila. 6: 175.
Elgaria prinicipis Baird and Girard, 1852, ibid.
Gerrhonotus formosus, O'Shaughnessy, 1873, Ann. Mag. Nat. Hist., ser. 4, 12: 47.
Gerrhonotus principis, Cope, 1875, Bull. U.S. Nat. Mus. 1: 46.
Gerrhonotus grandis, Yarrow, 1883, Bull. U.S. Nat. Mus. 24: 47. Part.
Gerrhonotus multicarinatus, Yarrow, 1883, ibid. Part.
Gerrhonotus scincicaudus, Yarrow, 1883, ibid., p. 48. Part.
Gerrhonotus palmeri, Van Denburgh, 1897, Occas. Pap. Calif. Acad. Sci. 5: 107.
Elgaria coerulea, Tihen, 1949, Amer. Midl. Nat. 41: 595.

Holotype. Zoologisches Museum, Berlin, 1163; Brazil, modified to San Francisco, California, by Stejneger (1902).

Derived features. Almost the only clear diagnostic feature of *E. coerulea* is its somewhat shortened, stocky body form, but this is not very helpful without comparative material. Three other derived features are seen in some specimens but are not useful for diagnosing the species as a whole: two lateral supraoculars, three suboculars, and three interoccipitals. The presence of dark eyes and ventral stripes between the scale rows may be diagnostic, but it is unclear whether these characters are ancestral or derived. In either case, they are useful in identification.

Color pattern. The color pattern of *E. coerulea* is extremely variable, ranging from a broad dorsal stripe flanked by darker barred or mottled sides to a condition in which 10-12 narrow to fairly broad dorsal crossbands are present; the dorsal ground color varies from bluish or brown to yellow or olive. The venter is either immaculate or has longitudinal stripes between the scale rows. The head and limbs are variable, from unmarked to heavily blotched and streaked with dark. The tail is similar to the dorsum in pattern.

Distribution. *Elgaria coerulea* is distributed from southern British Columbia south through western Washington and Oregon to central California and east to Idaho and Montana (Figure 28). It inhabits relatively cool, moist localities such as ocean beaches and clearings in northern and montane coniferous forests.

Discussion. This species is the most *Barisia/Mesaspis*-like species of *Elgaria*, and this led Stebbins (1958) to suggest relationship. Four subspecies (*coerulea*, *palmeri*, *principis*, and *shastensis*) are currently recognized (Fitch, 1938). Intergradation occurs among these forms.

Elgaria kingii Gray

Elgaria kingii Gray, 1838, Ann. Mag. Nat. Hist., ser. 1, 1: 390.
Gerrhonotus multifasciatus Duméril and Bibron, 1839, Erp. Gen. 5: 401.
Elgaria nobilis Baird and Girard, 1852, Proc. Acad. Nat. Sci. Phila. 6: 129.
Elgaria marginata Hallowell, 1852, Proc. Acad. Nat. Sci. Phila. 6: 179.
Gerrhonotus nobilis, Baird, 1859, U.S.-Mex. Bound. Surv. Rept., p. 11.
Gerrhonotus kingii, Müller, 1865, Reis. Ver. Staat. Can. Mex., p. 604.
Gerrhonotus knightii Herrick et al., 1899, Bull. Sci. Lab. Denison Univ. 11: 143.

Holotype. British Museum of Natural History, London, V25a/1946.8.29; Mexico, modified to Mojarachic, Chihuahua, Mexico, by Smith and Taylor (1950), and the Huachuca Mountains, Arizona, by Schmidt (1953).

Derived features. The single most outstanding diagnostic feature of *E. kingii* is its extreme elongation of body. Aside from this unique feature, the species is diagnosed by the presence of broken ventral stripes (yielding scattered spots). Several other derived characters are shared with various other *Elgaria* species; see the diagnosis of *E. cedrosensis* for most of these. Suffuse dark pigmentation between the dorsal crossbands is shared with *E. panamintina* and *E. parva*.

Color pattern. The color pattern of *E. kingii* is characterized primarily by a characteristic broad band of suffuse dark pigmentation anterior to each narrow dark dorsal crossband. The venter is pale, with spots lying on the scale rows. Characteristic black-and-white markings occur in the lateral fold and on the labials; otherwise the head is variously spotted. The tail is similar to the dorsum in color pattern and the limbs are variously mottled.

Distribution. *Elgaria kingii* is usually considered to be a species of relatively dry oak and oak-pine forests, although it is also known from semidesert. It is distributed from central Arizona and New Mexico south through western Mexico as far as Jalisco (Figure 28).

Discussion. Three subspecies have been described (*ferrea*, *kingii*, and *nobilis*).

Elgaria multicarinata (Blainville)

Cordylus multi-carinatus Blainville, 1835, Nouv. Ann. Mus. Nat. Hist. Natur. 4: 57.
Gerrhonotus multi-carinatus Blainville, 1835, ibid.: pl. 25.
Elgaria multicarinata, Gray, 1838, Ann. Mag. Nat. Hist., ser. 1, 1: 390.
Trachypeltis multicarinata, Fitzinger, 1843, Syst. Reptil., p. 21.
Gerrhonotus wiegmanni Gray, 1845, Cat. Spec. Liz. Coll. British Mus., p. 54.
Tropidolepis scincicauda Skilton, 1849, Amer. J. Sci. Arts 7: 202.
Elgaria grandis Baird and Girard, 1852, Proc. Acad. Nat. Sci. Phila. 6: 176.
Elgaria scincicauda, Baird and Girard, 1853, Stansb. Expl. Gt. Salt Lake, p. 348.
Gerrhonotus webbii Baird, 1858, Proc. Acad. Nat. Sci. Phila. 10: 255.
Gerrhonotus grandis, O'Shaughnessy, 1873, Ann. Mag. Nat. Hist., ser. 4, 12: 47.

Gerrhonotus scincicaudus, O'Shaughnessy, 1873, ibid.
Gerrhonotus caeruleus Boulenger, 1885, Cat. Liz. British Mus. 2: 273. Part.

Holotype. Museum National d'Histoire Naturelle, Paris, 2002; California, restricted to Monterey by Fitch (1938).

Derived features. See the diagnosis of *E. cedrosensis* for a list of derived features which *E. multicarinata* shares with other *Elgaria* species. No derived features unique to the species have been observed.

Color pattern. The color pattern of *E. multicarinata* consists basically of 10-12 relatively narrow dark dorsal crossbands on a background of brown, yellowish or reddish; the crossbands are usually distinct. The venter is paler, with longitudinal stripes running down the centers of the scale rows. The head is immaculate or variously spotted or blotched. The limbs are variously unmarked or mottled.

Distribution. *Elgaria multicarinata* inhabits relatively dry, warm areas (relative to *E. coerulea*) such as oak woodland, chaparral, and semidesert. It is distributed from southern Washington south through western Oregon and California to central Baja California, Mexico (Figure 28).

Discussion. *Elgaria multicarinata* intergrades among the five currently recognized subspecies (*ingava, multicarinata, nana, scincicauda,* and *webbii*; Fitch, 1938).

Elgaria panamintina (Stebbins)

Gerrhonotus panamintinus Stebbins, 1958, Am. Mus. Novitates 1883: 2.

Holotype. Museum of Vertebrate Zoology, Berkeley, 65410; Surprise Canyon, west side of Panamint Mountains, Inyo Co., California, 4500 ft.

Derived features. All of the derived features of *E. cedrosensis* are present in *E. panamintina* as well. In addition, suffuse pigmentation between the dorsal crossbands (seen also in *E. kingii* and *E. parva*) and loss of the black-and- white labial markings seen in several related *Elgaria* (see the diagnosis of *E. cedrosensis*) also occur in this species.

Color pattern. The color pattern of *E. panamintina* is similar to that of *E. kingii* except that the characteristic black-and-white markings on the labials in the latter species are lacking and the longitudinal ventral stripes are not broken.

Distribution. *Elgaria panamintina* is known only from the desert mountain ranges of eastern California (Figure 28), where it is probably restricted to the vicinity of canyon streams, although Banta (1963) suggested that it may be more wide-ranging in desert habitats.

Elgaria parva (Knight and Scudday)

Gerrhonotus parvus Knight and Scudday, 1985, Southwest. Nat. 30: 89-94.
Elgaria parva, Smith, 1986, Bull. Maryland Herp. Soc. 22: 21.

Holotype. Sul Ross State University, Alpine, 5538; 3 km SE Galeana, Nuevo León, Mexico.

Derived features. The outstanding diagnostic feature of *E. parva* is its miniaturized body form, being one of the smallest of gerrhonotines. In addition, a suite of derived scale characters diagnose the species: 11-12 infralabials, absence of keeling, loss of the triangular lower subocular characteristic of several related species, and a reversal to the ancestral patternless venter. Further derived features shared with other *Elgaria* species are listed in the diagnoses above.

Color pattern. Elgaria parva is similar in color pattern to *E. kingii* but without the ventral spotting characteristic of that species. The venter is immaculate. Labial markings are weak.

Distribution. This species is known from only two specimens, both collected among dead yucca leaves in the open pine forests surrounding Galeana, Nuevo León, Mexico (Figure 28).

Discussion. Knight and Scudday (1985) suggested a relationship between *E. parva* and *G. lugoi* on the basis of small size and reduced keeling. Both of these characters are derived but are also found in many other gerrhonotines, particularly in *Mesaspis*. Smith (1986) correctly pointed out that placement in *Elgaria* is more accurate, citing as *Elgaria* characters the lack of postrostrals, nasal-rostral contact, cantholoreal presence, anterior internasal absence, contact of the supranasals, and oviparity. All of these are characteristic of *Elgaria*, and nasal-rostral contact, anterior internasal absence, and contacting supranasals (which are all manifestations of anterior internasal loss; see above) are derived and unique to *Elgaria*.

Elgaria paucicarinata (Fitch)

Gerrhonotus multicarinatus, Dumeril and Bibron, 1839, Erp. Gen., p. 404. Part.
Gerrhonotus paucicarinatus Fitch, 1934, Copeia 1934: 7.
Elgaria paucicarinata, Tihen, 1949, Amer. Midl. Nat. 41: 595.

Holotype. Museum of Vertebrate Zoology, Berkeley, 11768; Todos Santos, Baja California Sur, Mexico.

Derived features. The diagnosis of *E. cedrosensis* above is equally applicable to this species; no unique derived characters have been observed.

Color pattern. The color pattern of *E. paucicarinata* is similar to that for *E. cedrosensis*, but the dorsal crossbands are usually more distinct and the ventral striping is often reduced. The tail is also more prominently banded.

Distribution. Although known primarily from the pine-oak forests of the mountains of the Cape region of Baja California (Figure 28), *E. paucicarinata* has also been collected in semidesert at near sea level.

Discussion. Almost the only real difference between this species and *E. cedrosensis* is distributional; only slight, probably overlapping, differences in color pattern otherwise differentiate them. Because of this similarity, their status as separate species is questionable and in fact Fitch (1938), who had originally described both forms (1934), admitted that the differences between them are at the level of subspecies in other *Elgaria*.

Bostic (1971) and Ottley (1983) suggested that *E. paucicarinata* and *E. multicarinata* should be considered conspecific. However, this analysis confirms that *E. paucicarinata* is more closely allied with *E. kingii*, as was suggested by Fitch (1938).

Barisia Gray

Gerrhonotus Wiegmann, 1828, Isis 21: 381. Part.
Barisia Gray, 1838, Ann. Mag. Nat. Hist., ser. 1, 1: 390.
Tropidogrrhonotus Fitzinger, 1843, Syst. Reptil., p. 21.
Barissia Gray, 1845, Cat. Spec. Liz. Coll. British Mus., p. 54.
Tropidogerrhum Agassiz, 1846, Nomencl. Zool., p. 203.

Type species. *Barisia imbricata* (Wiegmann).

Derived features. Of the derived gerrhonotine characters ancestral for *Barisia*, several have already been discussed as also occurring in *Gerrhonotus* and/or *Elgaria*. Certain other derived features are shared only with *Mesaspis* and *Abronia*: the presence of fewer than ten longitudinal dorsal rows between the hind limbs, eight or fewer nuchals, fewer than 100 caudal whorls and 15-17 scales per whorl, a stocky body form (seen also in *Elgaria coerulea*), and sexual dichromatism (at least in all species for which the condition of this character is known). Of the remaining derived features of *Barisia*, three are homoplastically shared with a few members of other genera: frontonasal absence (also in some *Mesaspis viridiflava*, *M. antauges*, and *Abronia oaxacae*), rugose postoccipitals (seen also in many *Abronia*), and 14 longitudinal dorsal rows (seen also in many *Gerrhonotus*, *Elgaria*, *Mesaspis*, and *Abronia*). Reduction to fewer than 40 transverse dorsal rows (also seen in *Abronia*) is also derived.

Four characters are unique to *Barisia* and join with the latter three above in diagnosing the genus: supranasal-postnasal fusion, canthal-cantholoreal fusion, a reduction from 5-7 to 1-3 superciliaries, and a characteristically short, deepened snout.

Discussion. Stejneger (1890) stated that, within *Barisia*, *B. imbricata* and *B. levicollis* are more closely allied than either is to *B. rudicollis*. He cited as evidence the presence in *B. rudicollis* of a low number of transverse dorsals, keeled nuchals, and nasal-rostral contact. Unfortunately, all of these characters are derived and therefore have no bearing on his argument. Tihen (1949b), in an analysis of the species here referred to the genera *Barisia* and *Mesaspis* (all of which he called *Barisia*), also suggested that within *Barisia* (in the sense of the present paper; his "*imbricata* group"), *B. rudicollis* is an early, although in some respects highly specialized, offshoot. The ancestral characters he cited were 14 longitudinal ventral rows, two loreals (=a loreal and a cantholoreal), four superciliaries instead of one or three, and extensive keeling. This last character is probably derived.

Barisia imbricata (Wiegmann)

Gerrhonotus imbricatus Wiegmann, 1828, Isis, 21: 381.
Gerrhonotus lichenigerus Wagler, 1830, Desc. Icon. Amph., fasc. 2, pl. 24, fig. 2.
Gerrhonotus adspersus Wiegmann, 1834, Herp. Mex., pl. 10.
Barisia imbricata, Gray, 1838, Ann. Mag. Nat. Hist., ser. 1, 1: 390.
Barisia lichenigera, Gray, 1838, ibid.
Barissia imbricata, Gray, 1845, Cat. Spec. Liz. Coll. British Mus., p. 55.
Barissia lichenigera, Gray, 1845, ibid.
Gerrhonotus olivaceus Baird, 1858, Proc. Acad. Nat. Sci. Phila. 10: 255.
Barissia olivacea, Cope, 1875, Bull. U.S. Nat. Mus. 1: 46.
Gerrhonotus planifrons Bocourt, 1879, Miss. Sci. Mex. Rept. 6: 361.
Barissia planifrons, Cope, 1887, Bull. U.S. Nat. Mus. 32: 1-98.
Gerrhonotus levicollis, Smith, 1942, Proc. U.S. Nat. Mus. 92: 368. Part.

Holotype. Zoologisches Museum, Berlin, 1156; Mexico; restricted to Distrito Federal, Mexico, by Smith and Taylor (1950).

Derived features. Among *Barisia* species, *B. imbricata* showed no uniquely derived features among those observed in the present study; both such are shared with *B. levicollis* (postnasal-anterior loreal fusion and almost complete loss of dorsal patterning).

Color pattern. Male *B. imbricata* are usually solid tan or brown with the head, venter, tail, and limbs immaculate. Females retain the ancestral condition, being usually darker than the male and having more prominent barring on the flanks.

Distribution. *Barisia imbricata* is one of the most widespread gerrhonotine species in Mexico, occurring in the highland pine forest areas throughout the country west of the Isthmus of Tehuantepec (Figure 29). It is usually found in clearings and other more-or-less open areas.

Discussion. As currently accepted, *B. imbricata* contains four subspecies (*ciliaris*, *imbricata*, *jonesi*, and *planifrons*), but see the discussion under *B. levicollis* below.

Barisia levicollis (Stejneger)

Barissia levicollis Stejneger, 1890, Proc. U.S. Nat. Mus. 13: 184.
Gerrhonotus imbricatus, Dunn, 1936, Proc. Acad. Nat. Sci. Phila. 88: 475. Part.
Gerrhonotus levicollis, Smith, 1942, Proc. U.S. Nat. Mus. 92: 368.
Barisia levicollis, Tihen, 1949, Am. Midl. Nat. 41: 598.

Holotype. United States National Museum, Washington, 9362; "Mexican boundary."

Derived features. Aside from the derived features shared with *B. imbricata* (see above), *B. levicollis* shows the following uniquely derived features: reduction of the superciliary row to one scale, more than 40 rows of transverse dorsals, and an increase to 16 longitudinal dorsal scale rows.

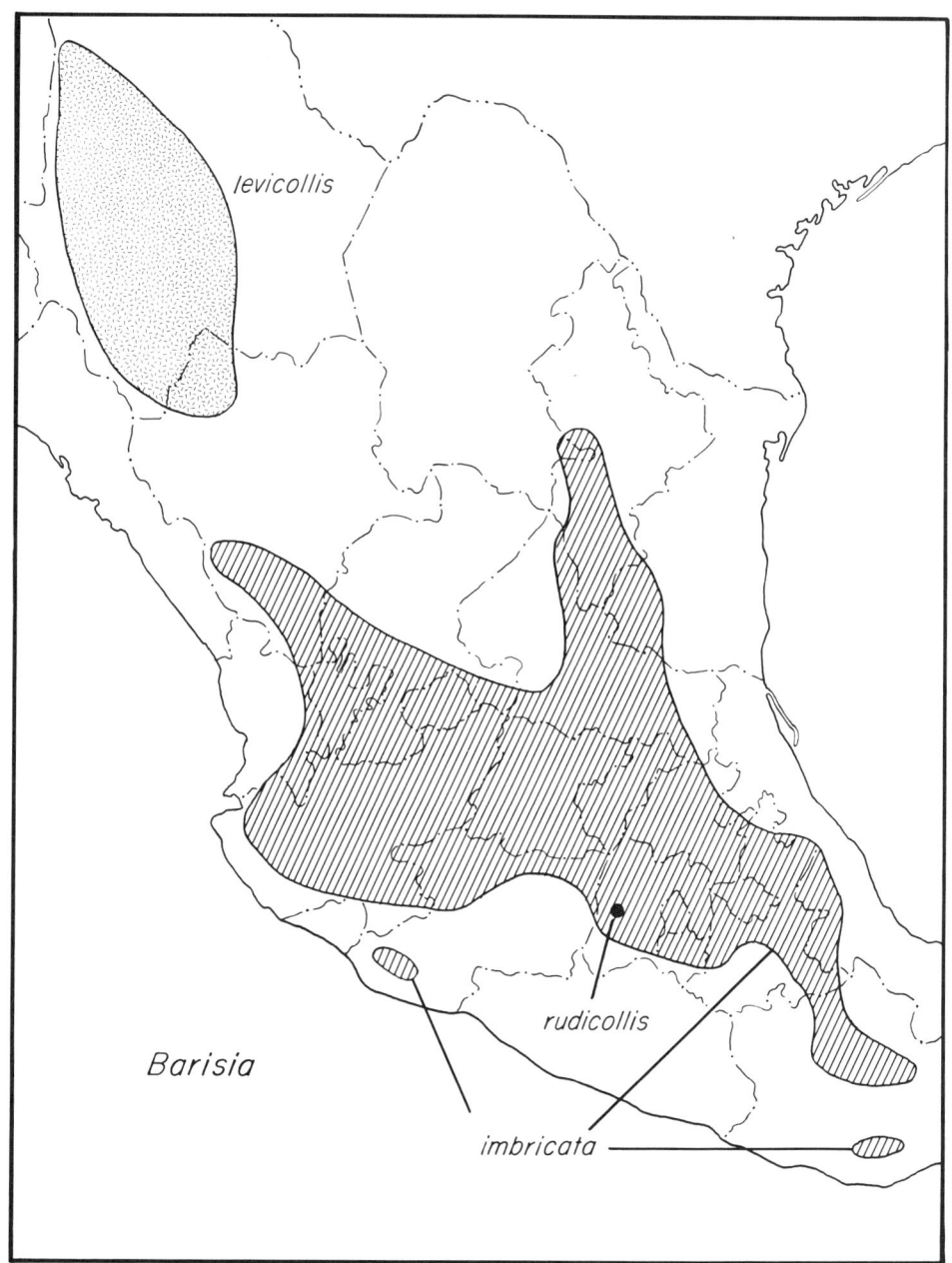

FIGURE 29. The geographic distribution of *Barisia*.

Color pattern. The color pattern of *B. levicollis* is similar to that of *B. imbricata*, except that both sexes tend to show the condition seen in *B. imbricata* males.

Distribution. *Barisia levicollis* inhabits similar upland openings to those occupied by *B. imbricata*. It occurs in the mountains of Chihuahua, Mexico (Figure 29).

Discussion. *Barisia levicollis* is probably no more different from *B. imbricata* than the various subspecies of *B. imbricata* are from each other; either its specific status or the subspecific status of some of the *B. imbricata* subspecies is therefore open to question. However, without further study, a conservative approach seems best: the species are presented here in their currently accepted taxonomy.

Barisia rudicollis (Wiegmann)

Gerrhonotus rudicollis Wiegmann, 1828, Isis, 21: 380.
Barisia rudicollis, Gray, 1838, Ann. Mag. Nat. Hist., ser. 1, 1: 390.
Tropidogrrhonotus rudicollis, Fitzinger, 1843, Syst. Reptil., p. 21.
Barissia rudicollis, Gray, 1845, Cat. Spec. Liz. Coll. British Mus., p. 55.
Tropidogerrhum rudicollis Agassiz, 1846, Nomencl. Zool., p. 203.

Holotype. Zoologisches Museum, Berlin, 1158; Mexico; restricted to Hacienda la Gavia, in the state of México, by Smith and Taylor (1950).

Derived features. *Barisia rudicollis* is the most derived of the three species of *Barisia*. Diagnostic features are: nasal-rostral contact, strongly keeled postoccipitals and very strong, almost acuminate keeling in general, 4-6 nuchals, 14 longitudinal ventrals, and relatively long, well-clawed limbs.

Color pattern. The color pattern of *B. rudicollis* in life is unknown. In alcohol, it is similar to that of *B. imbricata* except that the tail is more strongly banded and there is perhaps more pigmentation on the venter.

Distribution. The habitat of *B. rudicollis* is not certainly known. Although it apparently occurs in the pine forest regions of the state of México, Mexico (Figure 29), it is unclear whether it is terrestrial, as are the other species of *Barisia*, or arboreal, as suggested by its relatively elongate limbs and well developed claws. Wiegmann (1834) reported that *B. rudicollis* was "collected among rocks, much like skinks."

Discussion. *Barisia rudicollis*, although among the first six gerrhonotine species described (Wiegmann, 1828), still is known from only two specimens. Although certainly a member of the genus *Barisia*, it shows structural similarities to *Abronia* (elongate limbs, well developed claws, and a decreased nuchal number).

Mesaspis Cope

Barissia, Cope, 1866, Proc. Acad. Nat. Sci. Phila. 18: 132. Part.
Gerrhonotus, Bocourt, 1871, Bull. Nouv. Arch. Mus. 7: 102. Part.
Mesaspis Cope, 1878, Proc. Amer. Philos. Soc. 17: 96.
Barisia, Tihen, 1949, Amer. Midl. Nat. 41: 596. Part.

Type species. Mesaspis moreleti (Bocourt).

Derived features. Apart from the derived features discussed above in the diagnoses of *Gerrhonotus*, *Elgaria*, and *Barisia* as being diagnostic of more inclusive sets of genera, *Mesaspis* possesses five derived features. One of these, reduction of the lateral fold, occurs also in *Abronia* and in *Coloptychon*. The other four (eight rather than ten longitudinal ventrals at the forelimbs, subgranular scales on the leading edges of the shanks, and characteristic labial striping and ventral speckling) are restricted to *Mesaspis* and diagnose the genus.

Discussion. Tihen (1949b) divided what he saw as "*Barisia*" into three groups: the "*imbricata* group", here referred to as *Barisia*; the "*gadovii* group," including the forms *antauges*, *gadovii*, and *modesta*; and the "*moreleti* group," including *monticola*, *moreleti*, and *viridiflava*. The "*imbricata* group" (=*Barisia*) is discussed above. Of the other two groups, here included in the genus *Mesaspis*, Tihen's "*moreleti* group" is a monophyletic assemblage and is retained. The "*gadovii* group," however, is paraphyletic, since *M. antauges* (= *M. modesta*, see below) and *M. juarezi* together apparently form the sister taxon to the "*moreleti* group." I therefore recognize three monophyletic assemblages of species in *Mesaspis*: the *gadovii* group, containing the single species *M. gadovii*, and the *moreleti* and the *antauges* groups, containing *M. antauges* and *M. juarezi* and *M. viridiflava*, *M. monticola*, and *M. moreleti*, respectively.

I have elsewhere (Good, 1988b) analyzed allozyme variation among the species *M. gadovii*, *M. viridiflava*, *M. monticola*, and *M. moreleti* and have hypothesized relationships in concordance with the results of the present analysis: *M. monticola* and *M. moreleti* are sister taxa, *M. viridiflava* forms the sister taxon to them, and *M. gadovii* is the outgroup to all three. The *antauges* group species were not available for analysis.

The *gadovii* group

Derived features. As for the species below.

Mesaspis gadovii (Boulenger)

Gerrhonotus gadovii Boulenger, 1913, Ann. Mag. Nat. Hist., ser. 8, 12: 564.
Barisia gadovii, Tihen, 1949, Amer. Midl. Nat. 41: 598.

Syntypes. British Museum of Natural History, London, 1946.8.29.44-45, 1946.8.8.4-15; Omilteme, Guerrero, Mexico.

Derived features. *Mesaspis gadovii* is the only relatively large member of the genus (an ancestral character, but one which is useful in identification). Its only derived features are the presence of frontal-frontonasal contact in some specimens (seen also in some *M. monticola* and some *M. moreleti*) and the presence of a single subocular.

Color pattern. The color pattern of *M. gadovii* consists of a brown to reddish-brown dorsal stripe flanked by vertical bars on the sides, these latter sometimes accompanied by a white or yellow wash. The side bars sometimes extend onto the dorsum as chevron-shaped dorsal crossbands. The top of the head is similar to the dorsum, but the labials are strikingly colored with a white or yellow stripe bordered above by a black stripe. The venter is heavily spotted in males, less so in females. The tail is similar to the dorsum in pattern; the limbs are mottled.

Distribution. The habitat of *M. gadovii* is similar to that of *B. imbricata*, but the species is restricted to the highlands of the Sierra Madre del Sur of Guerrero and Oaxaca, Mexico (Figure 30).

Discussion. The two currently recognized subspecies (*gadovii* and *levigata*) differ in some minor characteristics of scalation and coloration (Spengler and Smith, 1982). The two are geographically isolated, but the minor structural differences do not warrant specific status.

The *antauges* group

Derived features. Among the derived *Mesaspis* characteristics seen in the *antauges* group, three are shared with all other species except *M. gadovii*: expansion of the supranasal, reduction in keeling, and reduction in body size. In addition, several are restricted to this group and are diagnostic: an enhancement of keeling reduction so that scale keels are virtually absent, presence of a postrostral, broad frontal- interparietal contact, and an elongate anterior superciliary.

Mesaspis antauges (Cope)

Barissia antauges Cope, 1866, Proc. Acad. Nat. Sci. Phila. 18: 132.
Pterogasterus modestus Cope, 1878, Proc. Amer. Philos. Soc. 17: 97.
Gerrhonotus modestus, Günther, 1885, Biol. Centr. Amer., Reptil., p. 42.
Gerrhonotus antauges, Sumichrast, 1882, La Naturaleza 6: 40.
Barisia angauges, Tihen, 1949, Am. Midl. Nat. 41: 598.
Barisia modesta, Tihen, 1949, ibid.

Holotype. United States National Museum, Washington, 30221; Mt. Orizaba, Veracruz, Mexico.

Derived features. Mesaspis antauges is differentiated from *M. juarezi* by color pattern and by a division of the posterior internasals into two scales, occasional partial fusion of the cantholoreal and the preocular, and extreme robustness of the male, all derived features.

Color pattern. The color pattern of *M. antauges* in life is unknown. In alcohol, the male is very dark brown, almost black, with a broad dorsal stripe flanked by darkly barred sides. The head is immaculate except for light coloration on the labials. The venter is dark with light flecking, the tail is similar to the dorsum in pattern and the limbs are immaculate or slightly mottled. The female color pattern is similar except that it is a much lighter brown and the side bars and ventral spotting are less distinct.

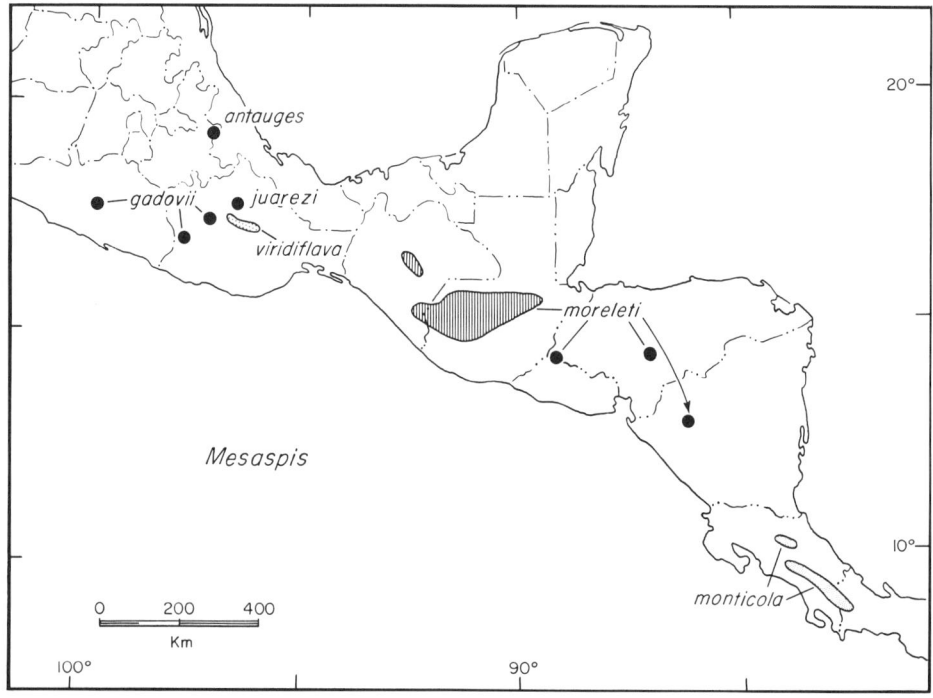

FIGURE 30. The geographic distribution of *Mesaspis*.

Distribution. Mesaspis antauges is known only from Mt. Orizaba, Veracruz, Mexico (Figure 30), but its exact distribution on the mountain is unknown. There is some indication that it may come from the bunchgrass regions above timberline (J. W. Wright, pers. comm.).

Discussion. Confusion exists over the status of the nominate species *M. modesta*, which was described by Cope (1878) from "Guatemala." Smith (1942) corrected the type locality to Mt. Orizaba, since the type specimens apparently bear labels to that effect (Tihen, 1949b). It is interesting to note that, in the two supposed species of *Mesaspis* occurring on Mt. Orizaba, all of the known specimens of "*M. modesta*" are female while all known specimens of *M. antauges* are male. The only differences between the two forms are in color pattern and in general robustness, *M. antauges* being much darker, much more solidly built, and broader of head. There are no scalational differences. Because sexual dimorphism is widespread in the genus, and because gerrhonotines are notorious for lack of intrageneric sympatry (see below), it seems clear that *M. antauges* and "*M. modesta*" represent the two sexes of a single highly dimorphic species.

Mesaspis juarezi (Karges and Wright)

Barisia juarezi Karges and Wright, 1987, Contr. Sci., Nat. Hist. Mus. Los Angeles Co. 381: 1.

Holotype. University of Texas, Arlington, R-8485; NE slope of Sierra Juárez between 6.1 and 11.6 km (3.8 and 7.2 mi) N crest of Cerro Pelón, Ixtlán District, Oaxaca, Mexico, 2500-2700 m.

Derived features. See *M. antauges* above.

Color pattern. The color pattern of *M. juarezi* is similar to that of *M. gadovii*, but without the dark markings on the side of the head.

Distribution. *Mesaspis juarezi* occurs on the forest floor and in clearings in cloud forest.

The *moreleti* group

Derived features. The *moreleti* group of *Mesaspis* is diagnosed primarily by the presence of a single postmental scale rather than the ancestral two. In addition, loss of one of the three ancestral lateral supraoculars and a high degree of canthal/loreal variability are diagnostic. A derived postnasal-anterior loreal fusion may also be ancestral for the group, but this is unclear.

Discussion. Stejneger (1907) suggested without explanation that *M. monticola* and *M. moreleti* are closely allied. Tihen (1949b), on the other hand, suggested that *M. moreleti* is probably the most primitive member of the *moreleti* group, on the basis of high a high number of dorsal scales, three lateral supraoculars instead of the two seen in *M. viridiflava* and *M. monticola*, a relatively high number of cantholoreal scales, and general coloration. A high number of dorsal scales is, in fact, ancestral for the Gerrhonotinae, but within *Mesaspis* it is probably derived. The high number of cantholoreal elements is shared with many *M. monticola* and is also probably derived in *Mesaspis*. The color pattern in *M. moreleti* is not considered here to be any more ancestral than that in the other two members of the group.

I have elsewhere (Good, 1988b) hypothesized on the basis of allozyme variation that *M. moreleti* and *M. monticola* are sister species and that *M. viridiflava* forms the outgroup to them.

Mesaspis monticola (Cope)

Gerrhonotus monticolus Cope, 1878, Proc. Amer. Philos. Soc. 17: 97.
Gerrhonotus alfaroi Stejneger, 1907, Proc. U.S. Nat. Mus. 32: 505.
Barisia monticola, Tihen, 1949, Am. Midl. Nat. 41: 598.

Holotype. United States National Museum, Washington, 30591; summit of Pico Blanco, Costa Rica.

Derived features. Within the *moreleti* group, *M. monticola* shares a number of derived features with *M. moreleti*: reduction or absence of posterior internasals (in many *M. moreleti*), midline expansion of the supranasals (in many *M. moreleti*), usually frontal-frontonasal contact (in almost all *M. moreleti*), the cantholoreal sometimes divided into a canthal and a loreal (in some *M. moreleti*), and usually anterior loreal-preocular contact (in most *M. moreleti*). Postnasal-anterior loreal fusion is shared with *M. viridiflava*. No derived scale features restricted to *M. monticola* have been detected.

Color pattern. The male in alcohol is dark brown to black, heavily spotted over the entire body (including the head, venter, tail, and limbs) with white. In life, bright yellow, green, and black predominate. The female shows the ancestral *Mesaspis* color pattern, being brown with a broad dorsal stripe and lacking the prominent light spotting of the male.

Distribution. *Mesaspis monticola* occurs in openings in and around pine and cloud forests in the highlands of Costa Rica and western Panama (Figure 30).

Mesaspis moreleti (Bocourt)

Gerrhonotus moreleti Bocourt, 1871, Bull. Nouv. Arch. Mus. 7: 102.
Gerrhonotus fulvus Bocourt, 1871, ibid., p. 104.
Mesaspis moreleti, Cope, 1878, Proc. Amer. Philos. Soc. 17: 96.
Mesaspis fulvus, Cope, 1878, ibid.
Gerrhonotus salvadorensis Schmidt, 1928, Field Mus. Nat. Hist., Zool. Ser., 12: 196.
Barisia moreleti, Tihen, 1949, Am. Midl. Nat. 41: 598.

Syntypes. Museum National d'Histoire Naturelle, Paris, 1188, 1267, 1268; "le Peten, ainsi que les forêts de pins de la Haute Vera-Paz" (Guatemala).

Derived features. The derived *moreleti* group features of *M. moreleti* that are shared with *M. monticola* are listed in the diagnosis of the latter species above. Diagnostic features restricted to *M. moreleti* are primarily correlated with a general increase in scale numbers: 18-20 longitudinal dorsal rows, ten dorsals at the hind limbs, and ten nuchals. Prefrontal-frontal fusion and transverse division of the preocular are also sometimes seen, but because of their relative rarity they are not diagnostic features of the species.

Color pattern. The color pattern of *M. moreleti* is dominated by a broad brown dorsal stripe, sometimes spotted or streaked with darker brown. Dark vertical bars occur on the sides. The head is immaculate or variously spotted, the venter is lighter than the dorsum, and often showing dark spots, the tail is similar to the dorsum and the limbs are usually immaculate.

Females are similar to the male in color pattern, except that the lateral barring and ventral spotting are less prominent.

Distribution. The habitat of *M. moreleti* is similar to that of *M. monticola*, but the species occurs in the highlands of Central America from the Isthmus of Tehuantepec, Mexico, to the Nicaraguan Depression (Figure 30).

Discussion. *Mesaspis moreleti* is perhaps the most variable species of gerrhonotine in its scale patterns, including those characters used to differentiate the five currently recognized subspecies (*fulva*, *moreleti*, *rafaeli*, *salvadorensis*, and *temporalis*). All of the diagnostic characters vary widely within these subspecies; their validity is therefore questionable.

Mesaspis viridiflava (Bocourt)

Gerrhonotus viridiflavus Bocourt, 1873, Ann. Sci. Nat., ser. 5, 17: unpaginated.
Gerrhonotus bocourti Peters, 1877, Monatsb. Akad. Wiss. Berlin 1876: 298.
Gerrhonotus antauges, Bocourt, 1878, Miss. Sci. Mex., Reptil. 5: 346. Part.
Gerrhonotus obscurus Günther, 1885, Biol. Centr. Amer., Reptil., p. 40.
Barisia viridiflava, Tihen, 1949, Am. Midl. Nat. 41: 598.

Holotype. Museum National d'Histoire Naturelle, Paris, 2920; Mexico; restricted to the highlands of central Oaxaca by Tihen (1949b).

Derived features. Within the *moreleti* group, *M. viridiflava* is diagnosed by the usually absent frontonasal, fusion of the anterior canthal with the cantholoreal, and the presence of 14 longitudinal dorsal scale rows. Longitudinal division of the anterior internasals into four scales is also often seen. Postnasal-anterior loreal fusion is shared with *M. monticola.*

Color pattern. In the male *M. viridiflava*, the color pattern consists of a broad brown dorsal stripe flanked by black, white-flecked sides. The top of the head is brown, the sides of the head and labials black-and-white, with the white primarily on the labials. The venter is black with white spots. The tail is similar in pattern to the dorsum and the limbs are immaculate or possess some black-and-white spotting.

The female is similar in pattern to the male but the strong black-and-white coloration of the venter, sides, and head is reduced.

Distribution. Mesaspis viridiflava occurs in and around pine forests in central Oaxaca, Mexico (Figure 30).

Abronia Gray

Gerrhonotus Wiegmann, 1828, Isis, 21: 379. Part.
Abronia Gray, 1838, Ann. Mag. Nat. Hist., ser. 1, 1: 389.
Aspidosoma Fitzinger, 1843, Syst. Reptil., p. 21.
Leiogerrhonotus Fitzinger, 1843, ibid.
Leiogerrhon Agassiz, 1846, Nomencl. Zool., p. 203.
Barissia, Cope, 1885, Proc. Amer. Philos. Soc. 22: 171. Part.

Type species: *Abronia deppii*(Wiegmann).

Derived features. *Abronia* species are stout-bodied, relatively long-limbed, short-tailed gerrhonotines with several features advantageous for life in their almost entirely arboreal habitat. *Abronia* as a genus shares a number of derived features with *Elgaria*, *Barisia*, and/or *Mesaspis*; see the diagnoses of those genera for discussions of these characters. In addition, *Abronia* shows the following derived gerrhonotine features, all of which diagnose the genus: loss of the fifth temporal row, reduction of the number of transverse dorsal rows to fewer than 40, reduction to six or fewer nuchal scales, loss of the lateral fold between the ear and the forelimbs, relatively long, well clawed limbs, a widened and depressed head, and usually a characteristic dorsal crossbanding pattern, at least in juvenile specimens. All but three of these characteristics are restricted to *Abronia*; the three that are not are all shared with *Barisia rudicollis*, which may be convergent on the arboreal habits of *Abronia* species: fewer than 40 transverse dorsals (seen also in *B. imbricata* and probably primitive for *Barisia*), six or fewer nuchals, and relatively long, well-clawed limbs.

Discussion. *Abronia* can be viewed as consisting of four monophyletic groups of species. One of these contains only *A. mitchelli* of the Sierra Madre del Sur of Oaxaca, Mexico. The second contains the other two "primitive" species (i.e., lacking derived features shared by all other *Abronia*): *A. reidi* of the Sierra de los Tuxtlas, Veracruz, and *A. ornelasi* of Cerro Baúl, Oaxaca. These two, here referred to as the "*reidi* group," may in fact form a monophyletic assemblage with *A. mitchelli*, although evidence for this is equivocal (see discussion above).

The third group corresponds to the "*aurita* group" of Tihen (1954) and contains all of the species of *Abronia* that are distributed east from the Isthmus of Tehuantepec, to El Salvador, except *A. ornelasi* and *A. bogerti*: (*A. aurita*, *A. lythrochila*, *A. matudai*, *A. montecristoi*, *A. ochoterenai*, *A. salvadorensis*, and *A. vasconcelosii*; *A. lythrochila*, *A. montecristoi*, and *A. salvadorensis* were unknown to Tihen).

Finally, Tihen's (1954) "*deppii* group" (not the same as Smith's 1942 "*deppii* group" within *Gerrhonotus* [sensu lato], which was equivalent to *Abronia* as here circumscribed) contains all of the remaining species and is distributed west and north from the Isthmus of Tehuantepec as far as Tamaulipas, Mexico, although it also includes *A. bogerti*, which occurs to the east of the Isthmus: *A. bogerti*, *A. chiszari*, *A. deppii*, *A. fuscolabialis*, *A. graminea*, *A. kalaina*, *A. mixteca*, *A. oaxacae*, and *A. taeniata*. *Abronia chiszari*, *A. kalaina*, and *A. mixteca* were unknown to Tihen.

As defined by Tihen, *A. mitchelli*, *A. ornelasi*, and *A. reidi* fall into the *deppii* group, which was, however, defined on the basis of ancestral characters (see above), so there is no phylogenetic evidence for affinity. In fact, the sharing of four derived characters (loss of subocular-temporal contact, reduction of the fourth temporal row, reduction in secondary temporal number, and an increased broadening of the head) suggests that the *aurita* group is more closely allied to the *deppii* group than are any of these three species.

The *mitchelli* group

Abronia mitchelli Campbell

Abronia mitchelli Campbell, 1982, Herpetologica 38: 356.

Holotype. University of Texas, Arlington, R-10000; Cerro Pelón, north slope of Sierra Juárez, Oaxaca, Mexico, 2750 m.

Derived features. Features of *A. mitchelli* which are derived within *Abronia* are its expanded supranasals, paired interoccipitals, one postoccipital row, four pairs of large chinshields, reduced dorsal osteoderms, and perhaps its apparent loss of dorsal crossbanding in the adult. The presence of 16 longitudinal dorsal rows may also be derived, but polarity is uncertain. Of these characters, all except the interoccipital and postoccipital numbers are shared convergently by various other *Abronia* species.

Color pattern. The color of *A. mitchelli* in life has been reported to be "gray-green mottled with black" (Campbell, 1982).

Distribution. *Abronia mitchelli* is known from a single specimen from the cloud forests of Cerro Pelón, Sierra Juárez, Oaxaca (Figure 31).

Discussion. Campbell (1982) suggested that *A. mitchelli* is close to *A. bogerti*, *A. chiszari*, and *A. reidi*, because all three posess a free canthal scale and because of the high number of transverse dorsals (34, 39, 39, and 34-36, respectively) in these species and the presence of 16 longitudinal dorsals in *A. mitchelli* and *A. chiszari*. The presence of a canthal and 16 longitudinal dorsals are both ancestral characters for the genus and the possession of 34 or more transverse dorsals is shared by a variety of other *Abronia* species; no derived features were presented to support Campbell's hypothesis.

The *reidi* group

Derived features. Diagnostic features of the *reidi* group include midline contact of the supranasals, contact of three primary temporals with the postoculars, and a characteristic pale edging to the dorsal scales. All of these characters are restricted to *reidi* group species. In addition, two derived characters may further diagnose the group, but intraspecific variation occurs and, with the small sample sizes available for both *A. reidi* and *A. ornelasi*, their status as useful synapomorphies is questionable. Both characters (frontal-frontonasal contact and canthal-supranasal fusion) are seen elsewhere in *Abronia*.

Abronia ornelasi Campbell

Abronia ornelasi Campbell, 1984, Herpetologica 40: 373.

Holotype. University of Texas, Arlington, R-6641; vicinity of Colonia Rodulfo Figueroa, Cerro Baúl, Oaxaca, (19 km NW Rizo de Oro, Chiapas), Mexico, ca. 1600 m.

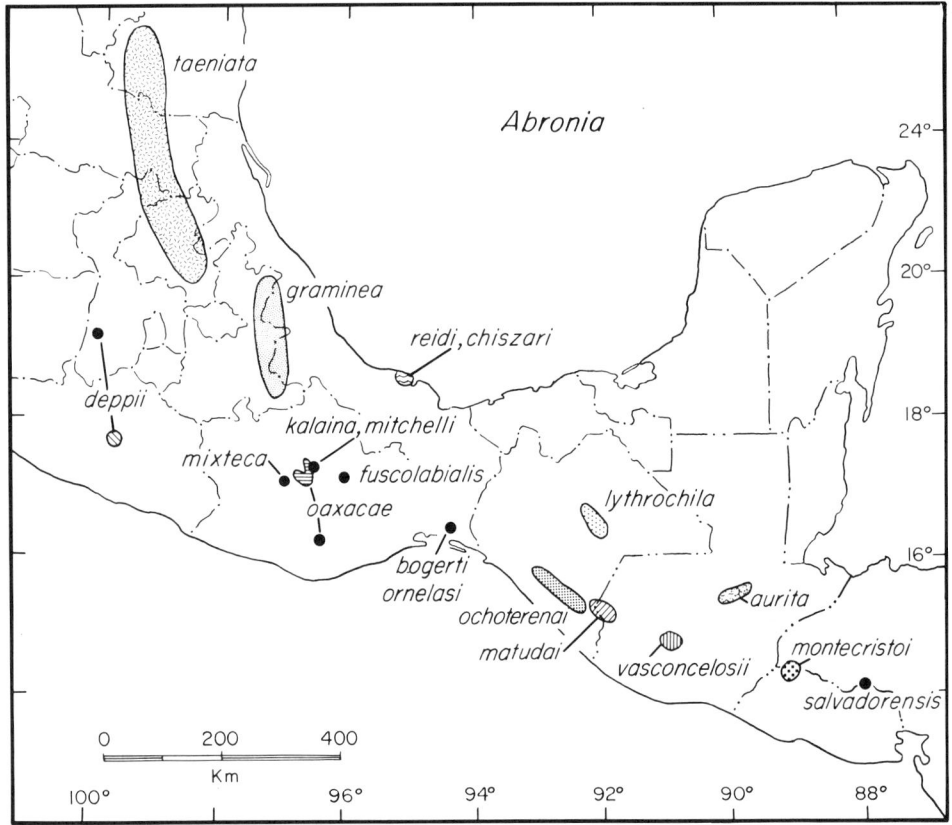

FIGURE 31. The geographic distribution of *Abronia*.

Derived features. No derived scale features of *A. ornelasi* unequivocally differentiate it from the other member of the *reidi* group, *A. reidi*. Only two such characters show any potential as diagnostic features for the species; both of these show variation in one or the other species and, since few specimens of either are known, little use can be made of them: frontal-frontonasal contact was seen in all *A. ornelasi* specimens examined and in one of the two known *A. reidi* and three suboculars occur occasionally in *A. ornelasi*.

Color pattern. In life (Campbell, 1984) the dorsum is solid brown with a slight greenish tint, each scale being edged posteriorly with light brown. The head is brown, the venter pale gray-green, and the tail brown above and plae gray-green below. The limbs are similar in color, except that the feet are yellow.

Distribution. Abronia ornelasi is known only from the vicinity of the type locality in the cloud forests of Cerro Baúl, Oaxaca, Mexico (Figure 31).

Discussion. Campbell (1984) proposed an alliance of *A. ornelasi* with *A. reidi* on the basis of the shared presence of midline contact of the supranasals, six nuchals, prominent dorsal keeling, 14 dorsal and 12 ventral longitudinal rows, and similar coloration. All of these but the first and last are ancestral within *Abronia* and do not provide evidence of relationships. The latter two, however, provide evidence and are among the synapomorphies of the *reidi* group.

Abronia reidi Werler and Shannon

Abronia reidi Werler and Shannon, 1961, Trans. Kansas Acad. Sci. 64: 123.

Holotype. University of Illinois Museum of Natural History, Urbana, 67062; crater rim, Volcán San Martín, Veracruz, Mexico, 5370 ft.

Derived features. Within the *reidi* group, *A. reidi* is diagnosed by prefrontal-superciliary contact, reduced dorsal osteoderm development, loss of the upper primary temporal, and probably supranasal-anterior canthal fusion (this latter may occur occasionally in *A. ornelasi*). Other possible diagnostic features include supranasal-rostral contact and posterior internasal-rostral contact, but both of these are variable in *A. reidi* and, since only two specimens of this species are known, are of limited usefulness.

Color pattern. The dorsum in life (Werler and Shannon, 1961) is dark olive-green with the posterior edges of the scales edged with pale yellow; this yellow coloration becomes predominant posteriorly. The top of the head is similar in color to the dorsum; the snout and labials are pale green. The temporals have broad lemon-yellow margins. The venter is immaculate and lighter in color than the dorsum. The tail and limbs are similar to the dorsum.

Distribution. The two known specimens of *A. reidi* were both collected from the cloud forests of Volcán San Martín, Sierra de los Tuxtlas, Veracruz, Mexico (Figure 31). They were collected 12-15 feet above the ground under moss on tree trunks (Werler and Shannon, 1961).

Discussion. Werler and Shannon (1961) detected a "lack of conformity" of *A. reidi* with either the *deppii* or *aurita* species groups (these groups as understood in 1961 corresponded to the same groups as discussed here; only after the inclusion of first *A. reidi* and later *A. mitchelli* and *A. ornelasi* in the *deppii* group was this correspondence lost). This "lack of conformity" is explained by the *deppii* group (sensu 1961) being defined on the basis of ancestral characters and that the *aurita* group is more closely allied with it than is *A. reidi*.

The *aurita* group

Derived features. The *aurita* group was defined by Tihen (1954) on the basis of two characters, the presence of a single postmental and protuberant supra-auricular scales. Both of these characters are derived and diagnose a monophyletic group among the species of *Abronia* known in 1954. However, *A. montecristoi* and *A. salvadorensis* were as yet unknown; they are here included in the *aurita* group, although they lack protuberant supra-auriculars.

Aside from the presence of a single postmental, the *aurita* group, as circumscribed here, is diagnosed by the presence of three subocular scales (seen also occasionally in *A.*

ornelasi) and loss of the posterior supralabials (somewhat intraspecifically variable in the group).

Abronia aurita (Cope)

Gerrhonotus auritus Cope, 1868, Proc. Acad. Nat. Sci. Phila. 20: 306.
Barissia fimbriata Cope, 1885, Proc. Amer. Philos. Soc. 22: 171.
Gerrhonotus fimbriatus, Günther, 1885, Biol. Cent. Am., Reptil., p. 37.

Holotype. United States National Museum, Washington, 6769; neighborhood of Petén and Cobán, Guatemala.

Derived features. Within the *aurita* group of *Abronia*, *A. aurita* possesses a number of derived features. Of these only one, the expansion of the second temporal row, causing a disruption of the third (without reduction in tertiary temporal number), is restricted to *A. aurita*. Other derived features include the presence of protuberent supra-auriculars and granular pre-auriculars, shared with all other *aurita* group members except *A. montecristoi* and *A. salvadorensis*, and 14 longitudinal ventral rows, seen in all *aurita* group species except *A. montecristoi*. Loss of the third primary temporal is shared with most *A. lythrochila* and *A. vasconcelosii*. Several intraspecifically variable characters may also be diagnostic of *A. aurita*, but small sample sizes make this determination impossible: secondary division of the single postmental (one postmental is a diagnostic feature of the *aurita* group) and frontal-frontonasal contact are seen in the *aurita* group only in some *A. aurita*. Loss of the third subocular (presence is another diagnostic characteristic of the *aurita* group) is seen in some *A. aurita* and in some *A. ochoterenai* and *A. vasconcelosii*.

Color pattern. The dorsum is light brown with ten darker crossbands. The head and limbs are also light brown while the venter and labials are off-white. The tail resembles the dorsum in color pattern.

Distribution. *Abronia aurita* occurs in the cloud forests of the Sierra de las Minas, central Guatemala (Figure 31).

Abronia lythrochila Smith and Alvarez del Toro

Abronia lythrochila Smith and Alvarez del Toro, 1963, Herpetologica 19: 100.

Holotype. University of Illinois Museum of Natural History, Urbana, 51013; Nachij, between Tuxtla Gutiérrez and San Cristóbal de las Casas, Chiapas, Mexico.

Derived features. Derived *aurita* group characters occurring in *A. lythrochila* include several shared with *A. aurita*, *A. matudai*, *A. ochoterenai*, *A. salvadorensis*, and/or *A. vasconcelosii*; these are listed in the diagnosis of *A. aurita* above. In addition, *A. lythrochila* shares a single character (an expanded posterior infralabial) only with *A. vasconcelosii*, and shows another which is seen only in most *A. lythrochila* (anterior canthal-posterior internasal fusion). The antepenultimate supralabial contacts the orbit in some *A. lythrochila* as well as in some *A. matudai* and *A. ochoterenai*, but the small samples available for these species makes this intraspecifically variable character of limited use.

Color pattern. Smith and Alvarez del Toro (1963) described *A. lythrochila* in life as being greenish-olive above with several irregular dark blotches, forming rings on the tail. Along the lateral fold, large sulfur-yellow blotches occur, as do reddish-brown blotches on the head, neck, and forepart of the trunk, between the dark bands. The infralabials are blood-red and the gular region is whitish, as are all other ventral surfaces. The anals are reddish-orange. Although not noted by Smith and Alvarez del Toro, some specimens show a hint of 10-11 dorsal crossbands. Color photographs were provided by Alvarez del Toro (1982).

Distribution. Abronia lythrochila occurs in the pine-oak forests of the Meseta Central of Chiapas, Mexico (Figure 31).

Abronia matudai (Hartweg and Tihen)

Gerrhonotus matudai Hartweg and Tihen, 1946, Occ. Pap. Mus. Zool., Univ. Michigan 497: 3.
Abronia matudai, Tihen, 1949, Amer. Midl. Nat. 41: 591.

Holotype. University of Michigan Museum of Zoology, Ann Arbor, 88831; Volcán Tacaná, Chiapas, Mexico, 2000 m.

Derived features. Several of the derived *aurita* group characters present in *A. matudai* are seen also in *A. aurita* and/or *A. lythrochila* and are listed above in the diagnoses of those species. Other derived features include 16 longitudinal dorsals and loss of the upper primary temporal (both restricted to *A. matudai*) and expanded lateral ventral rows (shared with *A. ochoterenai*). Two other intraspecifically variable characters may also diagnose the species but small sample sizes make this determination impossible: expanded supranasals (seen in the *aurita* group only in some *A. matudai*) and the return to the primitive *Abronia* condition of antepenultimate supralabial contact with the orbit (seen in some *A. matudai* and in some *A. ochoterenai* and some *A. lythrochila*).

Color pattern. The male is bright green. The female is light brown with 10-11 darker dorsal crossbands. The tail is similar to the dorsum in pattern and the head, venter, and limbs are immaculate or obscurely mottled.

Distribution. This species is known only from the cloud forests of Volcán Tacaná in southeastern Chiapas, Mexico, and Volcán Tajumulco in southwestern San Marcos, Guatemala (Figure 31).

Discussion. Hartweg and Tihen (1946) suggested that *A. vasconcelosii* might be the closest ally of *A. matudai* because of the occurrence in both of parietal-supraocular contact. This character is seen in a variety of *Abronia* species. Tihen (1954) suggested that *A. matudai* may be more closely allied with *A. bogerti* than with other members of the *aurita* group. However, this statement seems to have been made primarily on the basis of a relatively high number of transverse dorsal rows in both species, a character of dubious value within the genus.

Abronia montecristoi Hidalgo

Abronia montecristoi Hidalgo, 1983, Occ. Pap. Mus. Nat. Hist. Univ. Kansas 105: 6.

Holotype. University of Kansas Museum of Natural History, Lawrence, 184046; Hacienda Montecristo, Metapán, Cordillera de Alotepeque-Metapán, Dpto. de Santa Ana, El Salvador, 2250 m.

Derived features. Within the *aurita* group, *A. montecristoi* is diagnosed by a single derived feature: the presence of multiple interoccipitals. Variation occurs in the number of suboculars, but too few specimens of *A. montecristoi* are known for any claims about its status as a diagnostic feature to be made.

Color pattern. Wilson et al. (1986) describe the color pattern in life as "dorsum pale brown with indistinct brown crossbands; head horn-color; chin white; venter dirty white; palms of hands and feet yellowish-tan."

Distribution. *Abronia montecristoi* is known from only three specimens from the cloud forests of northern El Salvador and adjacent Honduras (Figure 31). It and *A. salvadorensis* represent the southernmost limit of the genus, the nearest congeneric being *A. vasconcelosii* from Volcán de Agua, southern Guatemala.

Abronia ochoterenai (Martín del Campo)

Gerrhonotus vasconcelosii ochoterenai Martín del Campo, 1939, An. Inst. Biol. Univ. Mexico 10: 357.
Gerrhonotus ochoterenai, Smith, 1942, Proc. U. S. Nat. Mus. 92: 368.
Abronia ochoterenai, Tihen, 1949, Amer. Midl. Nat. 41: 591.

Holotype. Instituto de Biología, Universidad Nacional Autonoma de México, Mexico City, 0338; Santa Rosa, Comitán, Chiapas, Mexico.

Derived features. *Abronia ochoterenai* possesses several derived features shared with the *aurita* group species described above; see the diagnoses of those species for lists. In addition, the upper primary temporal is sometimes divided in this species. Although this feature is restricted to *A. ochoterenai*, the small number of specimens available makes its use as a diagnostic feature impractical. No derived features unequivocally diagnose the species.

Color pattern. The color in males is bright green in life, duller in females (Smith and Alvarez del Toro, 1963). The dorsum, head, tail, and limbs are all more-or-less uniform in color or the dorsum shows 10-11 indistinct crossbands. The venter is lighter than the dorsum, and immaculate. Color photographs were provided by Alvarez del Toro (1982).

Distribution. This *Abronia* is known only from the cloud forests of the Sierra Madre de Chiapas, Mexico (Figure 31).

Discussion. Martín del Campo (1939) originally gave this form subspecific status within *A. vasconcelosii*, and thereby obviously suggested a relationship between the two.

He did not, however, discuss how the two forms were similar; only how they differed. The present analysis does not suggest sister taxon status for the two species.

Abronia salvadorensis Hidalgo

Abronia salvadorensis Hidalgo, 1983, Occ. Pap. Mus. Nat. Hist. Univ. Kansas 105:1.

Holotype. University of Kansas Museum of Natural History, Lawrence, 184047; Canton Palo Blanco, 10 km NE Perquín, Cordierra de Nahuaterique, Dpto. de Morazán, El Salvador, 1900 m.

Derived features. Lack of superciliary-cantholoreal contact, presence of four pairs of large chinshields and a reduction in number of dorsal crossbands diagnose *A. salvadorensis*. All of these features are unique among *aurita* group species.

Color pattern. Hidalgo (1983) described the color of *A. salvadorensis* in life as generally a grayish-cream background with five broad brown dorsal crossbands. The tail is similar to the dorsum in pattern. The head is grayish-cream, the venter cream, and the limb scales brown with cream edgings.

Distribution. A single specimen of *A. salvadorensis* is known, collected in the cloud forests of northeastern El Salvador (Figure 31).

Abronia vasconcelosii (Bocourt)

Gerrhonotus vasconcelosii Bocourt, 1871, Bull. Nouv. Arch. Mus. 7: 107.
Abronia vasconcelosii, Tihen, 1949, Am. Midl. Nat. 41: 591.

Holotype. Museum National d'Histoire Naturelle, Paris, 2017; Argueta, Guatemala.

Derived features. *Abronia vasconcelosii* shares several derived characters with various of the *aurita* group species listed above; see the diagnoses of those species. No derived features are restricted to the species.

Color pattern. The color pattern in life is unknown to me; in alcohol, the adult pattern is a solid light blue-gray without markings. The juvenile pattern shows 10 strong dorsal crossbands on a gray background and some dark flecking in the gular region.

Distribution. *Abronia vasconcelosii* is known only from south-central Guatemala (Figure 31). Its habitat is unknown to me; judging from the habitats occupied by its closest relatives (see above), it could occur either in cloud forest or in drier pine-oak forest.

The *deppii* group

Derived features. Tihen (1954) lumped the "*deppii* group" species together because they all lacked the two features of the *aurita* group (see above). Because it was defined solely by the lack of derived features, evidence for the monophyly of the group was lacking. However, two shared derived features not discussed by Tihen occur: nongranular neck scales and a gradual transition from granulars to nongranulars on the limbs.

Abronia bogerti Tihen

Abronia bogerti Tihen, 1954, Am. Mus. Novitates 1687: 3.

Holotype. American Museum of Natural History, New York, 68887; between Cerro Atravesado and Sierra Madre, Oaxaca, Mexico.

Derived features. Derived *deppii* group features seen in *A. bogerti* include longitudinal division of the anterior internasals, elongation of the anterior superciliaries, reduction of tertiary temporal number through expansion of the primary and secondary, expansion of the lower primary temporals at the expense of the upper so that only two scales are present, eight nuchal scales, and a characteristic light ventral flecking pattern. Of these, only the first is restricted to *A. bogerti*; the rest are shared with *A. chiszari* (primary temporals in this species are reduced to three scales rather than two).

Color pattern. The color pattern of *A. bogerti* in life is unknown. In alcohol (Tihen, 1954), the ground color is greenish, and 10-11 poorly defined brownish dorsal crossbands are present. The limbs and tail are similar in pattern to the dorsum. The venter shows scattered small black spots.

The only known specimen of *A. bogerti* is a juvenile, so the adult color pattern may differ somewhat.

Distribution. Abronia bogerti is known only from the type specimen collected in the southeastern Oaxacan highlands, Mexico (Campbell 1984) (Figure 31).

Discussion. Tihen (1954) saw *A. bogerti* as the "most generalized" species of *Abronia* then known. He viewed the presence of a relatively high number of transverse dorsals and eight rather than six nuchal scales as primitive, with the latter character being unknown elsewhere in the genus and the former being seen only in a single *aurita* group species (*A. matudai*). At the same time, he recognized that certain of the characters of *A. bogerti* are certainly derived (e.g., the temporal condition).

Both of the "generalized" characters of *A. bogerti*, although perhaps present in the outgroups to *Abronia*, are probably derived when viewed in the light of all characters combined. According to the present analysis, *Abronia bogerti* is an early offshoot of the *deppii* group, not basal to both the *deppii* and *aurita* groups.

Abronia chiszari Smith and Smith

Abronia chiszari Smith and Smith, 1981, Bull. Maryland Herpet. Soc.17: 51.

Holotype. University of Texas, Arlington, R-3195; 2.5 mi E Cuetzalapán, Sierra de los Tuxtlas, Veracruz, Mexico, 360 m.

Derived features. Abronia chiszari shares a number of derived *deppii* group features with *A. bogerti*; these are listed in the diagnosis of that species above. In addition, *A. chiszari* is the only *deppii* group species with 16 rather than the ancestral 14 longitudinal dorsal scales. The only known specimen of *A. chiszari* also shows prefrontal-anterior superciliary contact on one side and this may also be diagnostic of the species.

Color pattern. The color pattern of *A. chiszari* in life is unknown to me. In alcohol, the only known specimen (a juvenile) is light brown with 11 darker brown crossbands. The ventral surface is paler with small black flecks.

Distribution. The holotype of *A. chiszari* was collected on the bumper of a car in the lowland tropical forests of the Sierra de los Tuxtlas, Veracruz, Mexico (Figure 31).

Discussion. As stated by Smith and Smith (1981), *A. chiszari* resembles *A. bogerti* to such a degree that, if not for their difference in distribution, the two would be considered conspecific.

Abronia deppii (Wiegmann)

Gerrhonotus deppii Wiegmann, 1828, Isis, 21: 379.
Abronia deppii, Gray, 1838, Ann. Mag. Nat. Hist., ser. 1, 1: 389.
Leiogerrhonotus deppii, Fitzinger, 1843, Syst. Reptil., p. 21.
Gerrhonotus deppei Cope, 1868, Proc. Acad. Nat. Sci. Phila. 20: 306.

Holotype. Zoologisches Museum, Berlin, 1149; Mexico, restricted by Smith and Taylor (1950) to the vicinity of Omilteme, Guerrero, Mexico, but may instead be in the area of Temascaltepec and Real de Arriba in the state of México (Sanchez-Herrera and Lopez-Forment, 1980).

Derived features. Abronia deppii is one of the most highly derived *deppii* group species, with the following diagnostic characteristics: shared with all *deppii* group species except *A. bogerti* and *A. chiszari* are the presence of knob-like posterior head scales, a gradual transition from the lateral fold granulars to the dorsal scales, and an increase to 14 longitudinal ventral rows (this is sometimes reversed in *A. deppii*); shared with *A. graminea*, *A. mixteca*, *A. oaxacae*, and *A. taeniata* are contact of the nasal with the third supralabial, an increase in size of the posterior internasals, anterior canthal-posterior internasal fusion, rugose postoccipitals, and a reduction in the number of dorsal crossbands; shared with *A. mixteca* and *A. oaxacae* are enhanced knob-like posterior head scales, four pairs of large chinshields, 10-13 longitudinal dorsals, six longitudinal dorsals at the hind limbs, reduced dorsal keeling, posteromedially rounded flank scales yielding oblique lateral transverse dorsal rows, reduced dorsal osteoderms (seen also in *A. kalaina*), expanded neck scales, a strongly reduced lateral fold, and nongranular scales on the trailing edges of the limbs; shared with some *A. taeniata* is frontal-frontonasal contact (seen only occasionally in *A. deppii*); shared with *A. oaxacae* is a lack of superciliary-cantholoreal contact (seen only occasionally in *A. deppii*); shared with *A. mixteca* are three postoccipital rows and loss of dorsal osteoderms. In addition, *A. deppii* shows two unique derived features: a single subocular, and fusion of the lower primary temporal with the antepenultimate supralabial.

Color pattern. The dorsum of *A. deppii* is brown to gray with 6-8 darker crossbands. The tail is similar to to the dorsum in pattern and the head is usually unmarked. The venter is immaculate. The limbs are immaculate or lightly mottled.

Distribution. *Abronia deppii* is known from the pine-oak forests of central Guerrero, Mexico, and from similar habitat in the western part of the state of México, Mexico (Figure 31).

Abronia fuscolabialis (Tihen)

Gerrhonotus fuscolabialis Tihen, 1944, Copeia 1944: 112.
Abronia fuscolabialis, Tihen, 1949, Am. Midl. Nat. 41: 591.

Holotype. American Museum of Natural History, New York, 85634; Mt. Zempoaltepec, Oaxaca, Mexico.

Derived features. Several of the derived features of *A. fuscolabialis* are shared with other members of the *deppii* group and are listed in the diagnosis of *A. deppii* above. In addition to these, *A. fuscolabialis* shows three derived features, all of which are shared with *A. kalaina*: a gradual transition of ventrals to neck scales, contact of the penultimate supralabial with the orbit due to an anterior shift in the posterior margin of the mouth, and a characteristic ventral banding pattern. No diagnostic feature is restricted to *A. fuscolabialis* unless it is its bright green ground color.

Color pattern. The color pattern of *A. fuscolabialis* in life is a bright green background with 11-12 darker crossbands on the dorsum. The head, tail, and limbs are similarly mottled with darker pigmentation. The venter is lighter, with characteristic narrow crossbands on the posterior two thirds of the trunk and on the tail.

Distribution. *Abronia fuscolabialis* is known only from the vicinity of the type locality on Cerro Zempoaltepec, Oaxaca, Mexico (Figure 31).

Discussion. Tihen (1944, 1954) considered *A. fuscolabialis* to be most closely allied to *A. graminea*, but the only shared characters he discussed are ancestral among *Abronia*. Good and Schwenk (1985) allied *A. fuscolabialis* with *A. kalaina*, because both show a gradual transition from neck granulars to ventrals and characteristic ventral coloration, both derived features.

Abronia graminea (Cope)

Gerrhonotus gramineus Cope, 1864, Proc. Acad. Nat. Sci. Phila. 16: 179.
Abronia taeniata graminea, Tihen, 1949, Am. Midl. Nat. 41: 591.

Lectotype. United States National Museum, Washington, 6327; Orizaba, Veracruz, Mexico.

Derived features. Aside from the derived features shared with other *deppii* group species (see the diagnoses of *A. deppii* and *A. fuscolabialis*, above), *A. graminea* shares with *A. taeniata* the presence of granular preauriculars and a reversal to 12 longitudinal ventrals (this latter character is variable in *A. graminea*). *Abronia graminea* is distinguished from all other *deppii* group species including *A. taeniata* by the complete loss of dorsal patterning in the adult male and from all except *A. fuscolabialis* by bright green coloration. Some populations of *A. graminea* show the further diagnostic feature of having only four nuchal scale rows instead of the primitive six; this character is shared with *A. oaxacae*. Subocular-temporal contact is also intraspecifically variable in its derived presence in *A. graminea*. It is absent from all other *deppii* group *Abronia*.

Color pattern. The color pattern in the adult male consists of a bright green dorsum with no hint of crossbanding; the venter is yellow. Females are similar but duller, often with 6-8 faint crossbands.

Distribution. *Abronia graminea* occurs in the pine-oak forests in the highlands of central Veracruz and adjacent Puebla, Mexico (Figure 31).

Discussion. Tihen (1949a, 1954) combined the two forms *graminea* and *taeniata* into the single species *A. taeniata* on the basis of a collection of juvenile animals from La Joya, Veracruz, which he saw as intermediate in transverse dorsal row number, nuchal number, and color pattern. On examination of larger series, however, these animals fall within the range of *A. graminea* which is otherwise common at that locality. This population, although it is clearly *A. graminea* by adult color pattern, overlaps *A. taeniata* in these scale features. Although *A. taeniata* and *A. graminea* are completely allopatric and a test of their specific status through sympatry is impossible, this analysis agrees with the conclusion of Martin (1955) that there is no evidence that they should be combined; if Tihen hadn't proposed this association (on the basis of faulty evidence), specific status would never have been questioned.

Boulenger (1885) considered *A. taeniata* to be allied with *A. deppii* and *A. graminea* with *A. oaxacae*. He provided no evidence for this.

Abronia kalaina Good and Schwenk

Abronia kalaina Good and Schwenk, 1985, Copeia 1985: 135.

Holotype. Museum of Vertebrate Zoology, Berkeley, 177806; 16.6 km (by road) N summit on Hwy. 175, Cerro Pelón, Oaxaca, Mexico, 2300 m.

Derived features. See the diagnoses of *A. deppii* and *A. fuscolabialis* for lists of derived characters shared by *A. kalaina* and these forms; *A. kalaina* is most similar to *A. fuscolabialis*. The presence of midline frontoparietal contact, partial frontal-frontoparietal fusion, and midline contact of the second pair of chinshields are diagnostic features restricted to this species. Subocular-temporal contact is seen on one side of the only known specimen of *A. kalaina*, but there is no way to know whether this feature is diagnostic.

Color pattern. *Abronia kalaina* is similar in color pattern to *A. fuscolabialis*, except that the ground color is bright turquoise instead of bright green.

Distribution. This species is known only from the type locality in the cloud forests on Cerro Pelón, Oaxaca, Mexico (Figure 31).

Abronia mixteca Bogert and Porter

Abronia mixteca Bogert and Porter, 1967, Am. Mus. Novitates 2279: 2.

Holotype. American Museum of Natural History, New York, 91000; near Tejocotes, Oaxaca, Mexico, 2400 m.

Derived features. See the diagnosis of *A. deppii* for a list of the derived *deppii* group features shared by *A. mixteca* with most other species. In addition, *A. mixteca* possesses the following derived features: loss of the fourth temporal row, loss of the third primary temporal, three interoccipitals, and reduction in postmental size. Of these, the first two are variable in *A. mixteca*; all are shared with *A. oaxacae*, leaving *A. mixteca* with no unique derived features.

Color pattern. The dorsum of *A. mixteca* is gray to yellowish-olive with 6-8 indistinct crossbands, often reduced to dark blotches. The head and limbs are usually unmarked, the tail is similar in pattern to the dorsum, and the venter, sides of the neck, and labials are yellow.

Distribution. *Abronia mixteca* is known only from the vicinity of the type locality in the pine-oak forests near Tejocotes, Oaxaca, Mexico (Figure 31).

Discussion. Bogert and Porter (1967) suggested an alliance of *A. mixteca* with *A. oaxacae* because of the presence of three interoccipitals in both forms. This is a derived feature.

Abronia oaxacae (Günther)

Gerrhonotus oaxacae Günther, 1885, Biol. Cent. Am., Reptil., p. 36.
Abronia oaxacae, Tihen, 1949, Am. Midl. Nat. 41: 591.

Lectotype. British Museum of Natural History, London, 71.11.24.6; Oaxaca, Mexico.

Derived features. Aside from the shared derived features listed in the diagnoses of *A. deppii* and *A. mixteca* above, *A. oaxacae* possesses the following diagnostic characters: frontonasal reduction or absence, reduction in tertiary temporal number through increase in tertiary temporal size, loss of the posteriormost supralabial, and four nuchal scales (shared with some populations of *A. graminea*).

Color pattern. The dorsum of *A. oaxacae* is brownish or greenish usually with 6-8 dark crossbands. The head is brownish or greenish, the venter white, and the limbs and tail similar in pattern to the dorsum.

Distribution. Specimens of *A. oaxacae* are probably restricted to the highlands north and east of the city of Oaxaca, Mexico. Most collections have been made on Cerro San Felipe and on the south slope of the Sierra Juárez (Figure 31).

Abronia taeniata (Wiegmann)

Gerrhonotus taeniatus Wiegmann, 1828, Isis, 21: 379.
Abronia taeniata, Gray, 1838, Ann. Mag. Nat. Hist., ser. 1, 1: 390.
Aspidosoma taeniata, Fitzinger, 1843, Syst. Reptil., p. 21.
Gerrhonotus deppii, Mocquard, 1905, Bull. Mus. Natnl. Hist. Nat. Paris 1905: 79. Part.

Holotype. Zoologisches Museum, Berlin, 1152; Mexico; restricted to El Chico, Hidalgo, by Smith and Taylor (1950).

Derived features. All of the derived features of *A. taeniata* have been listed above in the diagnoses of the other *deppii* group species. No diagnostic scale features have been observed in *A. taeniata* that are not seen in at least one other species.

Color pattern. *Abronia taeniata* is essentially white, off-white or yellowish with 6-8 bluish-gray to black dorsal crossbands. Mottling is also present on the head, tail and limbs. The venter is immaculate.

Distribution. *Abronia taeniata* is the most widely distributed species in the genus. It occurs in pine-oak forest throughout much of eastern Mexico from southern Tamaulipas south to Hidalgo (Figure 31).

Discussion. Gehlbach and Collette (1957) suggested an alliance of *A. taeniata* with *A. oaxacae*. No evidence was provided.

THE TAXONOMIC IMPLICATIONS OF GERRHONOTINE RELATIONSHIPS

It is widely felt that taxonomy should reflect the evolutionary history of a group, and the only way this can be done is for taxonomic groups to be monophyletic. None of the recently suggested nomenclatural systems among the Gerrhonotinae (Tihen, 1949a, b, 1954; Stebbins 1958) fits this requirement. The frequent combining of *Gerrhonotus* and *Elgaria* into a single genus while retaining *Barisia* (sensu lato) is also untenable.

If this view that taxonomy should reflect phylogeny is taken, and the relationships suggested by this analysis accepted, a variety of approaches are possible, running the gamut from accepting a single all-inclusive genus (*Gerrhonotus*), as was done by most 20th-century workers before Tihen (1949a), to a separate genus for each species. This latter, of course, is ridiculous, but where one draws the line is not clear. Among gerrhonotines, the most prominent dichotomy is that between the long-limbed, long-clawed arboreal forms (*Abronia*) and the shorter-limbed terrestrial species (all other gerrhonotines, *Barisia rudicollis* may be an exception). It is generally thought that these arboreal species should be placed in their own genus, and this approach is followed here. If this is done, the division of the Gerrhonotinae into six genera is required: *Coloptychon*, *Gerrhonotus*, *Elgaria*, *Barisia*, *Mesaspis*, and *Abronia*. Any arrangement other than further subdivision of these genera would yield paraphyletic groups.

THE BIOGEOGRAPHIC IMPLICATIONS OF GERRHONOTINE RELATIONSHIPS

A robust hypothesis of phylogeny is necessary before any meaningful analyses of the geographic patterns in a group of organisms can be attempted; without such an hypothesis, inaccurate conclusions are inevitable. For example: it was generally accepted for many years that *Abronia* was the sister taxon to all other gerrhonotines, with the possible exception of *Coloptychon*. I have suggested (this analysis and Good, 1987b) that *Abronia* is instead the sister group only of the genus *Mesaspis*. Clearly, biogeographic hypotheses of the Gerrhonotinae based on these two views of relationship would be rather different.

Tihen (1949a) proposed a biogeographic history for the Gerrhonotinae based in part on what I consider as a mistaken view of relationships. In the present chapter, I discuss the biogeography of the subfamily from the point of view of the phylogeny presented in this analysis.

The Gerrhonotinae consists of four distinct biogeographic units: 1) the monotypic genus *Coloptychon* found in lowland Costa Rica and Panama; 2) *Gerrhonotus*, distributed primarily in Texas and through eastern and southern Mexico; 3) *Elgaria*, distributed throughout much of the western United States and western Mexico; and 4) a group of high-elevation species distributed throughout the mountainous regions of Middle America and contained in the genera *Barisia*, *Mesaspis*, and *Abronia*. I will first discuss the origin and diversification of the Gerrhonotinae and then briefly examine distribution patterns within them.

Savage (1966, 1982), presented the most in-depth discussions to date of the origins of the components of the Middle American herpetofauna. I will discuss the distribution of the Gerrhonotinae as it pertains to the hypotheses presented in those discussions.

THE ORIGIN AND EARLY DIVERSIFICATION OF THE GERRHONOTINAE

Savage (1966, 1982) viewed the pattern of distribution of components of the Central American herpetofauna as resulting from two major vicariant events. His scenario for the evolution of this fauna was as follows: prior to the early Tertiary, two major herpetofaunas existed in North America. All of Middle America (here defined as Mexico and Central

America), and as far north as at least 40º N, was occupied by a widespread tropical fauna which was contiguous with a similar fauna in South America. To the north, a basically circumpolar Laurasian fauna existed. Sometime in the early Tertiary, the original Isthmian Link between South and Central America was broken and the tropical faunas of the two continents began to diverge. The second vicariant event involved the isolation in Middle America of populations of northern derivation through mountain-building events and climatic changes in the Eocene-Miocene.

Savage saw the Central American herpetofauna as containing two elements (tracks) resulting from these events. The "Middle American element" consists of forms which arose from the original early Tertiary tropical fauna of Middle America. The "Old Northern element" is descended from the forms of northern ancestry which were isolated in the Eocene-Miocene and which evolved in proximity to the Middle American forms. Savage also saw two other elements as being important components of the Central American herpetofauna, both of them resulting from subsequent overcoming of the barriers discussed above: the South American element invaded Middle America following the re-establishment of the Isthmian Link in the late Pliocene, and the Young Northern element is a small group of relatively recent invaders from the drier regions of more northerly areas.

The Gerrhonotinae is, I think, a clear example of Savage's Old Northern element (however, see below). The sister group to the subfamily, as postulated I postulated elsewhere (Good, 1987a), is the Anguinae, a truly Laurasian group with extant members in eastern North America, Europe, the Middle East, East Asia, Sumatra, and Borneo. Since fossils apparently showing synapomorphies of the Anguinae (as well as most other anguid subfamilies) are known from the Eocene (Estes, 1983), a divergence of the gerrhonotine lineage from this group as a result of the second of Savage's vicariant events is not unreasonable. Savage (1966) listed the gerrhonotine genera as members of the Young Northern element, but corrected his view in his 1982 paper.

An alternate view of the possible origin of the Gerrhonotinae results if one accepts Gauthier's (1982) hypothesis that the Gerrhonotinae is the sister group to the Diploglossinae rather than to the Anguinae. A possible divergence of the Gerrhonotinae from the primarily South American Diploglossinae would then have occurred with the early Tertiary breaking of the Isthmian Link, thus making the Gerrhonotinae a member of the Middle American element. The inclusion of the Gerrhonotinae in this tropical assemblage is not beyond the realm of possibility since the sister group to all other gerrhonotines, *Coloptychon*, is the most southerly and the most "tropical" gerrhonotine, occurring in the lowlands of Costa Rica and Panama, immediately west of the former Panamanian Portal. However, because all of the anguid subfamilies, including a possible diploglossine, are known from the Eocene of North America (Gauthier, 1982), I prefer the hypothesis of northern origin and will proceed with this discussion on that assumption.

Either of the above scenarios makes a North American origin for the Gerrhonotinae virtually certain. Further, it is possible that the subfamily originated in Middle America, if the hypothesis that it represents an example of Savage's (1982) Old Northern Element is correct. In the latter case, the more northerly members (*Gerrhonotus* and *Elgaria* in the United States) probably represent more recent re-invasions of northern latitudes, either through active dispersal or possibly, in the case of the Pacific coast *Elgaria*, through rafting on drifting land masses (a similar pattern is seen among the primarily tropical American bolitoglossine salamanders, with *Batrachoseps* occurring in California and Oregon; D. B. Wake, pers. comm.).

Since fossil species which possess synapomorphies of the *Barisia/Mesaspis/Abronia* clade occurred in the Miocene and Pliocene (Good, 1988c), it is clear that divergence of the four major biogeographic subgroups of the Gerrhonotinae occurred before that time, although the exact sequence of events leading to this divergence is less clear. The presence of the outgroup to all other gerrhonotines, *Coloptychon*, at the southern limit of the range of the subfamily suggests one of two possibilities, assuming the Gerrhonotinae is accepted as being of northern derivation: either 1) the gerrhonotine ancestor to all of the present groups reached this southern limit (which corresponded to the southern limit of the North American continent at the time, the Panamanian Portal being open until sometime in the Pliocene) before radiating into the present forms, or 2) radiation occurred before this region was reached and *Coloptychon* later moved down into it and died out (or is yet undiscovered) in its area of origin. Lacking pertinent fossil material, I see no way to clearly distinguish between these two possibilities, although the former is more parsimonious as it requires a single vicariant event to form *Coloptychon*, while the latter requires three events: one leading to the formation of the genus, a dispersal event to get it to Costa Rica, and (probably) an extinction event in its area of origin.

THE BIOGEOGRAPHY OF *COLOPTYCHON*

Only a single species of *Coloptychon* is known, and the implications of its distribution are discussed above.

THE BIOGEOGRAPHY OF *GERRHONOTUS*

Gerrhonotus, as currently circumscribed, consists of two species. *Gerrhonotus liocephalus* occurs from the Edwards Plateau of central Texas south throughout low to mid elevations in eastern Mexico as far south as eastern Chiapas (and probably into Guatemala, although no specimens have yet been found). It reaches the Pacific slope of Mexico in Guerrero, Oaxaca, and Chiapas. Populations that are probably isolated occur in Chihuahua, Durango, Sinaloa, Jalisco, and possibly elsewhere.

Six subspecies are recognized in *Gerrhonotus liocephalus* but, although a definitive statement on this subject is beyond the scope of this work, it is likely that at least three of these should be considered distinct species. Each has a suite of morphological characters differentiating it from the others, and no intergradation is known. The subspecies *infernalis* is associated with the Edwards Plateau region of Texas and the moderate elevations of the Sierra Madre Oriental south to approximately the latitude of San Luis Potosí. South of this latitude, the subspecies *liocephalus* extends to the Isthmus of Tehuantepec, in similar habitats. The only form of *Gerrhonotus* to have invaded lowland habitats is the subspecies *ophiurus*, which occurs in the Gulf Coast lowlands of Veracruz.

The other three subspecies (*taylori*, *loweryi*, and *austrinus*) are less sharply defined, with each showing similarities to one of the above three.

Gerrhonotus is distributed primarily in the relative highlands of Mexico east of the currently uninhabited deserts of the Central Mexican Plateau. *Elgaria*, as will be discussed below, is distributed primarily west of this region. In each case, however, populations showing very close affinities to populations within these primary distributional areas occur on the other side of this barrier (e.g., *G. l. taylori* with *G. l. infernalis*, and *E. parva* with *E. kingii*). This suggests that the barrier is of relatively recent origin (Pleistocene?), and that the formation of this central Mexican desert was probably not responsible for the divergence of *Gerrhonotus* and *Elgaria*.

The second species of *Gerrhonotus*, *G. lugoi*, is known only from the Cuatro Ciénegas Basin of Coahuila, a general area also occupied by *G. liocephalus*. In this region, however, *G. lugoi* inhabits lower, more desert-like environments than *G. liocephalus*, which is restricted to higher elevations in the surrounding mountains. Morafka (1977) suggested that *G. lugoi* may be, in fact, merely the lower limit of an altitudinal cline in a variety of characters, at the other end of which is *G. liocephalus*.

THE BIOGEOGRAPHY OF *ELGARIA*

With the exception of *E. parva*, the biogeographic implications of which are discussed above, all of the species of *Elgaria* are distributed west of the central Mexican deserts.

The events leading to the divergence first of *Elgaria coerulea* and then of *E. multicarinata* from the rest of *Elgaria* cannot be accurately pinpointed, but it is likely that the Miocene-Pliocene tectonic events involved in the shifting of the San Andreas Fault and the rise of the Sierra Nevada in California, as described in detail by Yanev (1980), were involved. However unclear the divergence of these "early" *Elgaria*, the biogeographic history of the remainder of the genus (*E. kingii*, *E. panamintina*, *E. paucicarinata*, and *E. cedrosensis*) is much more certain. It is likely that the divergence of *E. panamintina*/ *E. kingii* from *E. paucicarinata*/*E. cedrosensis* (which are here considered sister species, although no synapomorphies were observed) is the result of a vicariant event associated with the late Miocene-early Pliocene separation of Baja California from mainland Mexico (cf. Murphy, 1983). Subsequent divergence of *E. panamintina* from *E. kingii* and *E. paucicarinata* from *E. cedrosensis* undoubtedly resulted from the development of intervening desert habitat in more recent (perhaps Pleistocene) times, as was suggested for the divergence of *E. multicarinata* and *E. paucicarinata* by Savage (1960).

THE BIOGEOGRAPHY OF *BARISIA*, *MESASPIS*, AND *ABRONIA*

The Middle American montane species of gerrhonotines form a monophyletic clade as illustrated in Figure 14. For the purposes of discussion, the montane areas they inhabit can be divided into the following regions (Figure 32): A) The northern Mexican highlands, consisting of the Sierra Madre Oriental, Sierra Madre Occidental, and Transverse Volcanic Belt. B) The Sierra Madre del Sur of Oaxaca and Guerrero which is separated from the

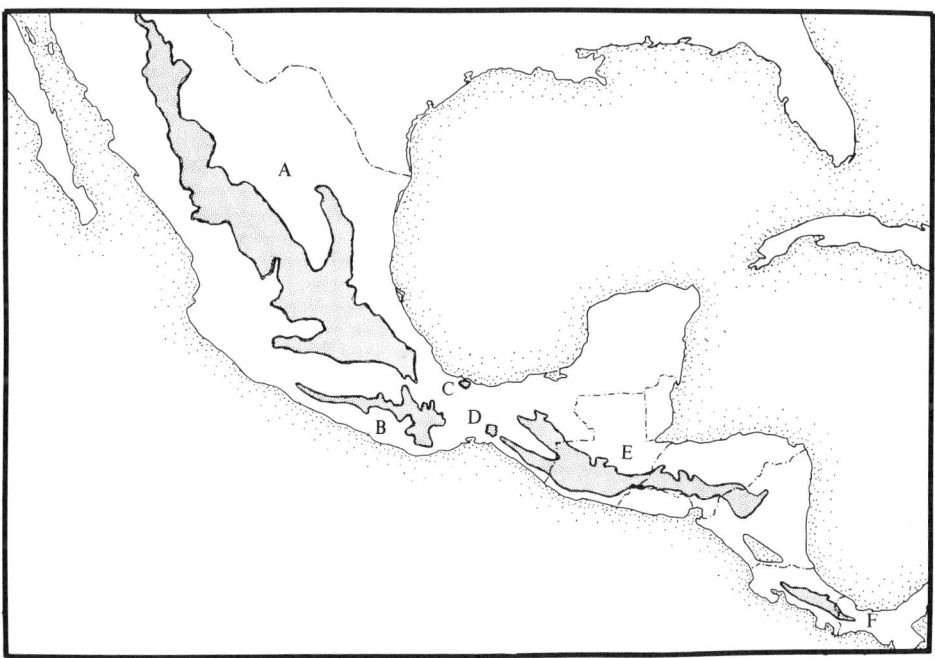

FIGURE 32. Major biogeographic regions based on the distribution of Middle American highland gerrhonotines. A=northern Mexican highlands (*Barisia imbricata, B. levicollis, B. rudicollis, Mesaspis antauges, Abronia graminea, A. taeniata, A. deppii*), B=Sierra Madre del Sur (*B. imbricata, M. gadovii, M. juarezi, M. viridiflava, A. mitchelli, A. kalaina, A. fuscolabialis, A. deppii, A. oaxacae, A. mixteca*), C=Sierra de los Tuxtlas (*A. reidi, A. chiszari*), D=southeast Oaxacan highlands (*A. ornelasi, A. bogerti*), E=Nuclear Central America (*M. moreleti, A. ochoterenai, A. matudai, A. aurita, A. lythrochila, A. vasconcelosii, A. salvadorensis, A. montecristoi*), F=Talamancan highlands (*M. monticola*).

northern Mexican highlands primarily by the valley of the Río Balsas. C) The Sierra de los Tuxtlas of coastal southern Veracruz; this highland area is isolated from the other ranges in southern Mexico by the low coastal plains of Veracruz. D) The southeastern Oaxacan highlands; an area of high elevation isolated from the Sierra Madre del Sur to the west by the Isthmus of Tehuantepec and from the Sierra Madre de Chiapas (one of the ranges of Nuclear Central America; see below) to the east by a series of low passes in the mountain chain that connects them. E) Nuclear Central America. These highlands extend from the Sierra Madre and Meseta Central de Chiapas southeastward to the Nicaraguan Depression, which isolates them from F) the Talamancan highlands of Costa Rica and western Panama. Herpetological aspects of the southern Mexican ranges (Areas B, C, D, and the western parts of E) were discussed in detail by Campbell (1984).

Twenty-seven species of high-elevation Middle American gerrhonotines are known. Savage (1982) provided area cladograms based on the herpetofauna as a whole for the six highland areas they inhabit. He showed that the Talamancan highlands are "related" to Nuclear Central America. Allied to these two areas is the Sierra Madre del Sur, and outside of all of these three are the highlands of northern Mexico. Gerrhonotines show almost exactly this pattern, with some modification.

Two of the three known species of *Barisia*, *B. rudicollis* and *B. levicollis*, occur only in the northern Mexican highlands; and the other, *B. imbricata*, occurs extensively there as well, although it also extends into the Sierra Madre del Sur. *Mesaspis antauges* is the only member of that genus to occur in the region. *Abronia taeniata*, *A. graminea*, and apparently a population of *A. deppii* (Sanchez-Herrera and Lopez-Forment, 1980) also occur there. In both *Mesaspis* and *Abronia*, the northern Mexican highland populations are closely allied with species occurring in the Sierra Madre del Sur and so are clearly not sister taxa to all other species. It seems likely, therefore, that the isolation of the northern Mexican highlands from the more southerly mountain ranges, possibly through the development of the Río Balsas basin and possibly through some earlier geologic or climatological event, was responsible for the vicariance of *Barisia* from *Mesaspis/Abronia*. The presence of *B. imbricata* in the Sierra Madre del Sur and of *M. antauges*, *A. graminea*, *A. taeniata*, and *A. deppii* in the northern Mexican highlands would then have to be the result of later dispersal across the intervening barrier and, at least in the case of *M. antauges* and *A. taeniata/A. graminea*, subsequent speciation.

Two explanations are possible for the distribution of *Mesaspis* and *Abronia* in the mountain ranges east of the Isthmus of Tehuantepec, assuming the presence of their common ancestor in the Sierra Madre del Sur, as required by the scenario above. Either they each dispersed eastward from a center of divergence in the Sierra Madre del Sur or they already inhabited the ranges on both sides of the Isthmus at the time of their divergence from each other, and gene flow continued across it until after this time.

In the case of *Mesaspis*, the former explanation seems more plausible, since only two members (*M. moreleti* and *M. monticola*) of a highly modified lineage, the *moreleti* group, occur east of the barrier, while all of the more generalized forms of *Mesaspis*, as well as the third member of the *moreleti* group (*M. viridiflava*), occur in the Sierra Madre del Sur.

The division within *Abronia* is more deep-seated. Excluding the "primitive" species *A. mitchelli*, *A. ornelasi*, and *A. reidi*, a major dichotomy exists between the *aurita* group of Nuclear Central America and the *deppii* group primarily of the Sierra Madre del Sur. The presence of these "primitive" *Abronia* east of the Isthmus in the southeast Oaxacan highlands (*A. ornelasi*), as well as west of it in the Sierra Madre del Sur (*A. mitchelli*) suggests that perhaps the proto-*Abronia* was more widespread than the proto-*Mesaspis*. Whatever the early history of the genus, invocation of dispersal would be necessary to account for the presence of *A. bogerti* east of the Isthmus in the southeast Oaxacan highlands, unless divergence of the *aurita* group and the *deppii* group resulted from some other cause prior to the development of a barrier at the Isthmus. Comparison with other taxa is of little help in this question, as there is great diversity in degree of divergence across this barrier (Campbell, 1984).

Two pairs of *Abronia* sister species show a particularly noteworthy pattern of distribution. *A. bogerti* and *A. ornelasi* both occur in the southeastern Oaxacan highlands and their respective sister species, *A. chiszari* (if such really exists; see above) and *A. reidi*, occur in the Sierra de los Tuxtlas of coastal Veracruz. This odd pattern of biogeographic relationship is shared with a variety of other amphibian and reptile species (Campbell, 1984).

Two fossil forms, apparently allied with *Abronia* (Good, 1988c), are known from the Pliocene of the United States (Estes, 1963; Holman, 1973). Their phylogenetic position either as sister taxa to modern *Abronia* species or within that genus is unclear, and it is therefore difficult to postulate a likely scenario explaining their presence there. If the former is the case, it is possible that the scenario presented above restricting the origin of *Abronia* in the mountains of southern Mexico and Central America is incorrect, and that it originated more to the north or at least in a wider area. If these fossil forms arose from within *Abronia*, dispersal northward in a period of favorable climate, and subsequent extinction, is possible. However, neither of these two possible phylogenetic positions precludes either scenario.

OTHER ASPECTS OF GERRHONOTINE BIOGEOGRAPHY

Allopatry

One of the most interesting aspects of the distribution of gerrhonotine species is their almost complete intrageneric allopatry. While sympatry among members of different genera is common, there are only two cases in which two members of a single genus are known to occur in the same area: *Barisia imbricata* and *B. rudicollis* in the western part of the state of México, Mexico, and *Elgaria coerulea* and *E. multicarinata* throughout much of northern California, Oregon, and southern Washington. In the former case, little is known about the habits of *B. rudicollis*; it may be arboreal and hence very different in microhabitat from *B. imbricata*. In the latter case, although *E. multicarinata* and *E. coerulea* are widely sympatric, they are largely allotopic, with *E. coerulea* occurring in cooler, moister habitats, as discussed below. Of the other suggested instances in which sympatry might potentially occur (*Abronia lythrochila* and *A. ochoterenai*, Smith and Alvarez del Toro, 1963; *A. chiszari* and *A. reidi*, Smith and Smith, 1981; *A. ornelasi* and *A. bogerti*, Campbell, 1984; *A. kalaina* and *A. mitchelli*, Good and Schwenk, 1985), no records yet exist.

Habitat

There is a dichotomy among gerrhonotines in the general habitat type occupied. *Gerrhonotus*, *Elgaria* (except *E. coerulea*), and *Coloptychon* inhabit low to moderate elevations and relatively warm habitats (relative to other gerrhonotines; it must be remembered that these do not always approach the warmth of habitats occupied by many other lizards). *Barisia*, *Mesaspis*, *Abronia*, and *E. coerulea* (which are also the only live-bearing gerrhonotines; Smith et al., 1983) inhabit high-elevation and/or cool, moist situations. Clearly, the most parsimonious explanation for this distribution of habitat utilization is that the lower, warmer habitat is ancestral and the higher, cooler habitat derived, and in fact this is in keeping with the ancestral habitat type occupied by each of the other anguid subfamilies. However, it must be recognized that the primitive state for

Elgaria is not discernible from this analysis, since the only cool habitat species, *E. coerulea*, is the sister group to all others. It is therefore equally parsimonious to explain this as a parallel occupation of cool habitats in this species and in the Middle American highland forms, or an occupation of such habitat by the common ancestor of these genera and *Elgaria* and a reversal to the primitive habitat in all species of *Elgaria* but *E. coerulea*.

Appendix A

Specimens Examined

The following is a list of the alcohol-preserved specimens examined in the course of this analysis. The number in parentheses following each species name is the total number of specimens examined. Museum acronyms are as follows:

AMNH, American Museum of Natural History, New York, NY.
ASU, Arizona State University, Tempe, AZ.
BMNH, British Museum of Natural History, London, UK.
CAS, California Academy of Sciences, San Francisco, CA.
FMNH, Field Museum of Natural History, Chicago, IL.
KU, University of Kansas, Lawrence, KS.
LACM, Natural History Museum of Los Angeles County, Los Angeles, CA.
MCZ, Museum of Comparative Zoology, Harvard University, Cambridge, MA.
MNHP, Museum National de l'Histoire Naturelle, Paris, France.
MVZ, Museum of Vertebrate Zoology, University of California, Berkeley, CA.
SRSU, Sul Ross State University, Alpine, TX.
UCR, Universidad de Costa Rica, San José, Costa Rica.
UIMNH, University of Illinois Museum of Natural History, Urbana, IL.
UMMZ, University of Michigan Museum of Zoology, Ann Arbor, MI.
UNAM, Instituto de Biología, Universidad Autonoma de México, Mexico City, Mexico.
USNM, United States National Museum, Washington, DC.
UTACV, University of Texas at Arlington, Collection of Vertebrates, Arlington, TX.
ZMB, Zoologisches Museum, Berlin, Germany.

Abronia aurita (4): MVZ 143461, 144537, 160608-609.
Abronia bogerti (1): AMNH 68887 (holotype).
Abronia chiszari (1): UTACV R-3195 (holotype).
Abronia deppii (18): CAS 143109; FMNH 105600; LACM 109262, 127415; MCZ 33750, 42716-717, 85248; MVZ 45005, 57172-173, 110941, 134106, 164922; USNM 113172, 148889-891.
Abronia fuscolabialis (2): LACM 15132; UTACV R-9899.

Appendix A

Abronia graminea (35): FMNH 71002; LACM 7547, 17704, 67703, 75507, 75509-510, 109263, 109959, 122489-493, 135558; MVZ 57465-466, 106322-325, 106763-764, 111213, 130000, 137081, 146941-943, 191065-070.
Abronia kalaina (1): MVZ 177806 (holotype).
Abronia lythrochila (4): CAS 141906; LACM 130124; MVZ 57170; UTACV R-3354.
Abronia matudai (4): LACM 75514; MVZ 161022, 161793; UMMZ 88331 (holotype).
Abronia mitchelli (1): UTACV R-10000 (holotype).
Abronia mixteca (8): AMNH 102640-643, 102647; LACM 121914, 121916, 125364.
Abronia montecristoi (1): KU 184046 (holotype).
Abronia oaxacae (10): LACM 122482-488; MVZ 144197, 164364; UIMNH 48672.
Abronia ochoterenai (2): UIMNH 52085-086.
Abronia ornelasi (3): UTACV R-6075, 6219, 7710.
Abronia reidi (1): UIMNH 73732.
Abronia salvadorensis (1): KU 184047 (holotype).
Abronia taeniata (12): LACM 15159, 106749-750, 109264; MVZ 109492-493, 128981, 191071-075.
Abronia vasconcelosii (2): MNHP 2017 (holotype); UMMZ 129013.

Barisia imbricata (32): FMNH 1504, 6526-528, 105762, 105765, 105769, 105775-777, 106165; MVZ 31714-717, 67399, 68943-946, 104145, 106326, 109494-496, 140649-650, 186485-486, 186714, 191048-049.
Barisia levicollis (20): FMNH 15728-729, 112021; MCZ 6977; MVZ 66067-068, 68782-783, 70699-701, 72857-858, 84638-639; USNM 26608, 26612, 46666-667, 47413.
Barisia rudicollis (1): UNAM 2701.

Coloptychon rhombifer (3): UCR 3143, 6971; ZMB 8655 (holotype; photograph and drawings seen).

Elgaria cedrosensis (7): CAS 56187 (holotype); FMNH 130594-598; MCZ 45730.
Elgaria coerulea: several hundred MVZ specimens.
Elgaria kingii (20): MVZ 13846, 42585, 49913, 56316, 61785, 64100, 68781, 69076, 70262, 84510, 104729, 149967-970, 191077-078, 193583-585.
Elgaria multicarinata: several hundred MVZ specimens.
Elgaria panamintina (16): MVZ 65403-409, 65410 (holotype), 65411, 77063, 134111, 150326-329, 191076.
Elgaria parva (1): SRSU 5538 (holotype), drawings and photographs seen.
Elgaria paucicarinata (10): MVZ 11768 (holotype), 45367-369, 50078-079, 191079-082.

Gerrhonotus liocephalus (15): MVZ 10323, 18942, 21227-229, 24847, 25361, 39667, 45006-007, 128079, 150322-324; UMMZ 94921.
Gerrhonotus lugoi (1): ASU 8818.

Mesaspis antauges (3): BMNH 1903.9.30.122; CAS 98681; UNAM 3685.
Mesaspis gadovii (37): FMNH 114604-618; MVZ 45008-017, 57174-176, 112388, 112393-394, 144198-199, 162300-302, 164778.

Mesaspis monticola (19): MVZ 78728-730, 79956-958, 92522, 97841-843, 113716-717, 128686, 191062-064, 193574-576.
Mesaspis moreleti: 275 MVZ specimens.
Mesaspis viridiflava (25): MVZ 112389-392, 140643-646, 162291-299, 191051-057, 191059.

Appendix B

Characters Used in the Analysis of Gerrhonotine Relationships

In this list, numbers refer to characters and letters to character states. Several of these general characters are subdivided into multiple characters in Appendix C and Tables 1-5. Ancestral/derived designations for character states in among-genera and within-genera analyses may differ (see text for explanation); consult Appendix C and the tables for individual analyses.

1. Nasal-rostral contact without loss of anterior internasals
    a. Present
    b. Absent or contact due to loss of anterior internasals
2. Supranasal-rostral contact
    a. Present
    b. Absent
3. Postrostral number
    a. Zero
    b. One
    c. Two
4. Nasal-third supralabial contact
    a. Present
    b. Absent
5. Anterior internasals
    a. Present
    b. Absent
6. Longitudinal division of anterior internasals
    a. Present
    b. Absent
7. Posterior internasal size
    a. Approximately twice the size of the anterior internasals
    b. Approximately equal to the anterior internasals in size
    c. Reduced or absent

8. Posterior internasal division
    a. Present
    b. Absent
9. Posterior internasal-rostral contact
    a. Present
    b. Absent
10. Supranasal expansion without loss of anterior internasals
    a. None or expansion due to loss of anterior internasals
    b. Expanded, but not to the midline
    c. Midline contact (unless excluded by postrostrals)
11. Supranasal-postnasal fusion
    a. Present
    b. Absent
12. Supranasal-frontonasal contact without posterior internasal reduction
    a. Present
    b. Absent or contact due to reduction of the posterior internasals
13. Postnasal-anterior loreal fusion
    a. Present
    b. Absent
14. Relative postnasal size
    a. Lower much larger than upper
    b. Approximately equal
15. Frontonasal
    a. Present
    b. Absent
16. Frontonasal-frontal contact
    a. Present
    b. Absent
17. Canthal/loreal number
    a. Five or more
    b. Four or fewer
    c. Three or fewer
18. Cantholoreal-anterior canthal fusion
    a. Present
    b. Absent
19. Anterior canthal-posterior internasal fusion
    a. Present
    b. Absent
20. Anterior loreal-preocular contact
    a. Present
    b. Absent
21. Partial fusion of cantholoreal and preocular
    a. Present
    b. Absent

Appendix B

22. High degree of canthal/loreal variability
     a. Present
     b. Absent
23. Prefrontal-superciliary contact
     a. Present
     b. Absent
24. Prefrontal-frontal fusion
     a. Present
     b. Absent
25. Frontal-interparietal contact
     a. Present
     b. Absent
26. Frontal-frontoparietal fusion
     a. Present
     b. Absent
27. Frontal-interparietal contact width
     a. Broad
     b. Narrow
28. Transverse division of frontal into two scales
     a. Present
     b. Absent
29. Sunken appearance of frontal
     a. Present
     b. Absent
30. Lateral supraocular number
     a. Two
     b. Three
     c. Four or five
31. Superciliary number
     a. More than three
     b. Three
     c. One
32. Superciliary-cantholoreal contact
     a. Present
     b. Absent, or lost due to reduction in superciliary number
33. Elongate anterior superciliary
     a. Present
     b. Not elongate
34. Preocular number
     a. More than one
     b. One
35. Transverse division of preocular
     a. Present
     b. Absent

36. Subocular number
    a. Four or more
    b. Three
    c. Two
    d. One
37. Subocular differentiation from pre- and postoculars
    a. Well differentiated
    b. Poorly differentiated
38. Subocular-temporal contact
    a. Present
    b. Absent
39. Triangular-shaped lower subocular
    a. Present
    b. Absent
40. Number of temporals per row
    a. Five or more
    b. Fewer than five
41. Fifth temporal row
    a. Present
    b. Absent
42. Fourth temporal row
    a. Present
    b. Reduced
    c. Absent
43. Reduction of tertiary temporal number through expansion of temporal rows one and two
    a. Present
    b. Absent
44. Reduction of tertiary temporal number through increase in tertiary temporal size
    a. Present
    b. Absent
45. Disruption of third temporal row by expansion of the second without reduction in tertiary temporal number
    a. Present
    b. Absent
46. Reduction in secondary temporal number
    a. Present
    b. Absent
47. Division of upper primary temporal
    a. Present
    b. Absent
48. Loss of upper (fourth) primary temporal
    a. Present
    b. Absent

49. Loss of third primary temporal
    a. Present
    b. Absent
50. Lowest (first) primary-penultimate supralabial fusion
    a. Present
    b. Absent
51. Expansion of lower primary temporals at expense of upper
    a. Present
    b. Absent
52. Number of primary temporals contacting orbit
    a. Three
    b. Fewer than three
53. Parietal-upper tertiary temporal contact
    a. Present
    b. Absent
54. Interoccipital number
    a. One
    b. Two
    c. Three
    d. Five
55. Number of postoccipital rows
    a. One
    b. Two
    c. Three
56. Postoccipital scale surfaces
    a. Smooth
    b. Rugose
    c. Strongly keeled
57. Knob-like posterior head scales
    a. Present
    b. Enhanced
    c. Absent
58. Protuberent supra-auriculars
    a. Present
    b. Absent
59. Pre-auricular size
    a. Granular
    b. Nongranular
60. Supralabial number
    a. 9-12
    b. 13-14
61. Loss of posterior supralabials
    a. Present
    b. Absent

62. Anterior shift of posterior margin of mouth
    a. Present
    b. Absent
63. Infralabial number
    a. 8-10
    b. 11-12
64. Expansion of posteriormost infralabial
    a. Present
    b. Absent
65. Postmental number
    a. One
    b. Two
66. Reduction in postmental size
    a. Present
    b. Absent
67. Sublabial number
    a. 6-8
    b. 4-5
68. Chinshield number/size
    a. Three large and one small
    b. Four large
    c. Four large and one small
    d. Five large
69. Midline contact of second pair of chinshields
    a. Present
    b. Absent
70. Transverse dorsal number
    a. More than 40
    b. Fewer than 40
71. Longitudinal dorsal number
    a. 18-20
    b. 16
    c. 14
    d. 10-13
72. Number of longitudinal dorsals at hind limbs
    a. Ten
    b. Eight
    c. Six
73. Keeling strength
    a. Very strong
    b. Strong
    c. Reduced
    d. Absent, except when the result of miniaturization of body size

74. Posteromedially rounded flank scales yielding oblique lateral transverse dorsal rows
    a. Present
    b. Absent
75. Dorsal osteoderm development
    a. Strong
    b. Reduced
    c. Absent
76. Number of longitudinal nuchal rows
    a. Twelve
    b. Ten
    c. Eight
    d. Six
    e. Four
77. Neck scale size
    a. Granular
    b. Subgranular
    c. Larger than subgranular
78. Gradual transition of neck scales to ventrals
    a. Present
    b. Absent
79. Lateral fold strength
    a. Unreduced
    b. Slight reduction
    c. Moderate reduction
    d. Strong reduction
80. Lateral fold between ear and forelimb
    a. Present
    b. Absent
81. Gradual transition from lateral fold granulars to dorsals
    a. Present
    b. Absent
82. Longitudinal ventral row number
    a. 10
    b. 12, without expansion of lateral rows
    c. 14, or 12 with lateral rows expanded
83. Expansion of lateral longitudinal ventral rows
    a. Present
    b. Absent
84. Number of longitudinal ventral rows between forelimbs
    a. Ten
    b. Eight
85. Size of scales on trailing edge of limbs
    a. Granular
    b. Nongranular

86. Gradual transition granulars-nongranulars on limbs
    a. Present
    b. Absent
87. Subgranulars on leading edge of shank
    a. Present
    b. Absent
88. Caudal whorl number
    a. More than 100
    b. Fewer than 100
89. Number of scales in whorls at base of tail
    a. 15-17
    b. 20-24
90. Body form
    a. Long, slender
    b. Short, stocky
91. Body size
    a. Large
    b. Small
92. Long, well clawed limbs
    a. Present
    b. Absent
93. Widened, depressed head
    a. Absent
    b. Moderate
    c. Pronounced
94. Short, deep snout
    a. Present
    b. Absent
95. Adult color pattern
    a. Dorsal stripe
    b. *Coloptychon*-like crossbanding
    c. *Gerrhonotus*-like crossbanding
    d. *Elgaria*-like crossbanding
    e. *Abronia*-like crossbanding
96. Suffuse, dark pigmentation between dorsal cross bands
    a. Present
    b. Absent
97. Number of crossbands
    a. 6-8
    b. More than nine
98. Characteristic light coloration on posterior edge of scales
    a. Present
    b. Absent

99. Loss of dorsal patterning in the adult
    a. Present
    b. Absent
100. Ventral markings
    a. None
    b. Chevron-like bands on throat and anterior part of trunk
    c. Longitudinal stripes between scale rows
    d. Longitudinal stripes on scale rows
    e. Broken longitudinal stripes
    f. Characteristic ventral speckling
    g. Crossbands on posterior part of trunk and on tail
    h. Characteristic scattered light flecking
101. Black-and-white markings in lateral fold
    a. Present
    b. Absent
102. Head spotting
    a. Present
    b. Absent
103. Black and white labial markings
    a. Present
    b. Absent
104. Characteristic labial striping
    a. Present
    b. Absent
105. Sexual dichromatism
    a. Present
    b. Absent

Appendix C

Distribution of Derived and Ancestral Character States Among the Genera of Gerrhonotine Lizards

In the list below, characters are designated by numbers corresponding to those in Appendix B. Some of the general characters in Appendix B are subdivided into multiple characters here (designated 3-1, 3-2, etc.). Character states are designated by letters corresponding to those in Appendix B. Polarity is established by outgroup comparison with the Diploglossinae and Anguinae. When polarity is ambiguous, letter designations are given to alternative states. 0=ancestral, 1=derived, N=not known. Col=*Coloptychon*, Ger=*Gerrhonotus*, Elg=*Elgaria*, Bar=*Barisia*, Mes=*Mesaspis*, Abr=*Abronia*.

Char.	Ancestral state (0)	Derived state (1)	Col	Ger	Elg	Bar	Mes	Abr
1	b	a	0	0	0	0/1	0	0
2	b	a	0	0	0	0	0	0/1
3-1	a/b	c	0/1	0	0	0	0	0
3-2	a/b	a/b	a	a/b	a	a	a/b	a
4	b	a	1	0/1	0	0	0	0/1
5	a	b	0	0	1	0	0	0
6	b	a	0	0	0	0	0/1	0/1
7-1	b	a	0	0	0	0	0	0/1
7-2	b	c	0	0	0	0	0/1	0
8	b	a	0	0	0	0	0/1	0
9	b	a	0	0	0	0	0	0/1
10-1	a	b/c	1	0/1	0	0	0/1	0/1
10-2	a/b	c	1	0	0	0	0/1	0/1
11	b	a	0	0	0	1	0	0
12	b	a	0	0	0	0	0	0/1
13	b	a	0	0	0	0/1	0/1	0
14	b	a	0/1	0/1	0	0	0	0
15	a	b	0	0	0	1	0/1	0/1

Appendix C

Char.	Ancestral state (0)	Derived state (1)	Col	Ger	Elg	Bar	Mes	Abr
16	a	b	1	1	1	1	0/1	0/1
17-1	a	b/c	0	0/1	1	1	1	1
17-2	a/b	c	0	0/1	1	1	0/1	1
18	b	a	0	0	0	0/1	0/1	0
19	b	a	0	0	0	0	0	0/1
20	b	a	0	0	0	0	0/1	0
21	b	a	0	0	0	0	0/1	0
22	b	a	0	0	0/1	0	0/1	0
23	b	a	1	0/1	0	0	0	0/1
24	b	a	0	0	0	0	0/1	0
25	a	b	0	0	0	0	0	0/1
26	b	a	0	0	0	0	0	0/1
27	b	a	0	0	0	0	0/1	0
28	b	a	0/1	0	0	0	0	0
29	b	a	0	0	0	0/1	0	0
30-1	b	a	0	0	0/1	0	0/1	0
30-2	b	c	0/1	0	0	0	0	0
31-1	a	b/c	0	0	0	1	0	0
31-2	a/b	c	0	0	0	0/1	0	0
32	a	b	0	0	0	N	0	0/1
33	b	a	0	0	0	0	0/1	0/1
34	a/b	a/b	a	a/b	b	b	b	b
35	b	a	0	0	0	0	0/1	0
36-1	a/b/c	a/b/c	a/b	b/c	b/c	c	c	b/c
36-2	a/b/c	d	0	0	0	0	0/1	0/1
37	a/b	a/b	b	b	a	a	a	a
38	a	b	0	0	0/1	0	0	0/1
39	b	a	0	0	0/1	0	0	0
40	a	b	0	0/1	1	1	1	1
41	a	b	0	0	0	0	0	1
42-1	a	b/c	0	0	0	0	0	0/1
42-2	a/b	c	0	0	0	0	0	0/1
43	b	a	0	0	0	0	0	0/1
44	b	a	0	0	0	0	0	0/1
45	b	a	0	0	0	0	0	0/1
46	b	a	0	0	0	0	0	0/1
47	b	a	0	0	0	0	0	0/1
48	b	a	0	0	0	0	0	0/1
49	b	a	0	0	0	0	0	0/1
50	b	a	0	0	0	0	0	0/1

Appendix C

Char.	Ancestral state (0)	Derived state (1)	Col	Ger	Elg	Bar	Mes	Abr
51	b	a	0	0	0	0	0	0/1
52	a	b	0	0/1	1	1	1	0/1
53	b	a	0/1	0	0	0	0	0
54-1	a	b	0	0	0	0	0	0/1
54-2	a	c	0	0	0/1	0	0	0/1
54-3	a	d	0	0	0	0	0	0/1
55-1	b	a	0	0	0	0	0	0/1
55-2	b	c	0	0	0	0	0	0/1
56-1	b	a	0	0	0	1	0	0/1
56-2	b	c	0	0	0	0/1	0	0
57-1	c	a/b	0	0	0	0	0	0/1
57-2	a/c	b	0	0	0	0	0	0/1
58	b	a	0	0	0	0	0	0/1
59	b	a	0	0	0	0	0	0/1
60	a/b	a/b	b	b	a	a	a	a
61	b	a	0	0	0	0	0	0/1
62	b	a	0	0	0	0	0	0/1
63	a/b	a/b	a	a	a/b	b	b	b
64	b	a	0	0	0	0	0	0/1
65	a	b	1	1	1	1	0/1	0/1
66	b	a	0	0	0	0	0	0/1
67	a/b	a/b	a	a	b	b	b	b
68-1	a	b	0	0	0	0	0	0/1
68-2	a	c	0	1	0	0	0	0
68-3	a	d	1	N	0	0	0	0
69	b	a	0	0	0	0	0	0/1
70	a	b	0	0	0	0/1	0	0/1
71-1	b	a	0	0/1	0	0	0/1	0
71-2	b	c/d	0	0/1	0	0/1	0/1	0/1
71-3	b/c	d	0	0	0	0	0	0/1
72-1	a	b/c	0	0	0	1	0/1	0/1
72-2	a/b	c	0	0	0	0	0	0/1
73-1	d	a/b/c	0	1	1	1	1	1
73-2	b/d	c	0	0	0/1	0/N	0	0/1
73-3	b/d	a	0	0	0	0/1	0	0
74	b	a	0	0	0	0	0	0/1
75-1	a	b/c	1	0	0	0	0	0/1
75-2	a/b	c	0	0	0	0	0	0/1
76-1	a	b/c/d/e	0	1	1	1	1	1
76-2	a/b	c/d/e	0	0	0	1	0/1	1

Appendix C

Char.	Ancestral state (0)	Derived state (1)	Col	Ger	Elg	Bar	Mes	Abr
76-3	a/b/c	d/e	0	0	0	0/1	0	0/1
76-4	a/b/c/d	e	0	0	0	0/1	0	0/1
77-1	a	b/c	0	0	0	0	0	0/1
77-2	a/b	c	0	0	0	0	0	0/1
78	b	a	0	0	0	0	0	0/1
79-1	a	b/c/d	1	0	0	0	1	1
79-2	a/b	c/d	0	0	0	0	1	1
79-3	a/b/c	d	0	0	0	0	0	0/1
80	a	b	0	0	0	0	0	1
81	b	a	0	0	0	0	0	0/1
82-1	a	b/c	0	1	1	1	1	1
82-2	a/b	c	0	0/1	0	0/1	0	0/1
83	b	a	0	0	0	0	0	0/1
84	a	b	0	0	0	0	1	0
85	b	a	0	0	1	1	1	0/1
86	a/b	a/b	a	a	a	a	a	a/b
87	b	a	0	0	0	0	1	0
88	a/b	a/b	a	a	a	b	b	b
89	b	a	0	0	0	1	1	1
90	a	b	0	0	0/1	1	1	1
91	a	b	0	0/1	0/1	0	0/1	0
92	b	a	0	0	0	0/1	0	1
93-1	a	b/c	0	0	0	0	0	1
93-2	a/b	c	0	0	0	0	0	0/1
94	b	a	0	0	0	1	0	0
95-1	a	b	1	0	0	0	0	0
95-2	a	c	0	1	0	0	0	0
95-3	a	d	0	0	0/1	0	0	0
95-4	a	e	0	0	0	0	0	1
96	a/b	a/b	a	a	a/b	a	a	a
97	a/b	a/b	b	b	b	b	b	a/b
98	b	a	0	0	0	0	0	0/1
99	b	a	0	0	0	0/1	0	0/1
100-1	a	b	1	0	0	0	0	0
100-2	a	c	0	0	0/1	0	0	0
100-3	a	d	0	0	0/1	0	0	0
100-4	a	e	0	0	0/1	0	0	0
100-5	a	f	0	0	0	0	1	0
100-6	a	g	0	0	0	0	0	0/1
100-7	a	h	0	0	0	0	0	0/1

Char.	Ancestral state (0)	Derived state (1)	Col	Ger	Elg	Bar	Mes	Abr
101	b	a	0	0	0/1	0	0	0
102	b	a	0	0	0/1	0	0/1	0
103	b	a	0	0	0/1	0	0	0
104	b	a	0	0	0	0	1	0
105	b	a	0	0	0	1	1	1

Literature Cited

Agassiz, L.
1846. Nomenclatoris Zoologici Index Universalis. Jent and Gassmann, Soloduri. 393 pp.

Alvarez del Toro, M.
1982. Los Reptiles de Chiapas. Inst. Historia Natural, Tuxtla Gutierrez, Chiapas, Mexico. 247 pp.

Anderson, J. D., and W. Z. Lidicker
1963. A contribution to our knowledge of the herpetofauna of the Mexican state of Aguascalientes. Herpetologica 19: 40-51.

Baird, S. F.
1858. Description of new genera and species of North American lizards in the museum of the Smithsonian Institution. Proc. Acad. Nat. Sci. Phila. 10: 253-256.
1859. Reptiles of the Boundary, with notes by the naturalists of the survey. U. S.-Mex. Boundary Surv. 3: 1-35.

Baird, S. F., and C. Girard
1852. Description of new species of reptiles, collected by the U. S. Exploring Expedition under the command of Capt. Charles Wilkes, U.S.N. Proc. Acad. Nat. Sci. Phila. 6: 125-129.
1853. Reptiles, *in* H. Stansbury. Exploration and survey of the valley of the Great Salt Lake of Utah, pp. 336-365. Special Session, U. S. Senate.

Banta, B. H.
1963. Remarks upon the natural history of *Gerrhonotus panamintinus* Stebbins. Occ. Pap. Calif. Acad. Sci. 36: 1-12.

Blainville, M. H. D. de
1835.   Description de quelques espèces de reptiles de la Californie. Nouv. Ann. Mus. Nat. Hist. Natur. 4: 1-64.

Bocourt, M.-F.
1871.   Description de quelques Gerrhonotes nouveaux provenant de Mexique et de l'Amérique Centrale. Bull. Nouv. Arch. Mus. 7: 101-108.
1873.   Notes erpétologiques. Ann. Sci. Nat., ser. 5, 17: unpaginated.
1878.   In A. M. C. Duméril, G. Bibron, and M. F. Mocquard. Études sur les Reptiles, Mission Scientifique au Mexique et dans l'Amérique Centrale--Recherches Zoologiques, vol. 5. Imprimerie Impériale, Paris. pp. 346-349
1879.   In A. M. C. Duméril, G. Bibron, and M. F. Mocquard. Études sur les Reptiles, Mission Scientifique au Mexique et dans l'Amérique Centrale--Recherches Zoologiques, vol. 6. Imprimerie Impériale, Paris. pp. 361-363

Bogert, C. M., and A. P. Porter
1967.   A new species of *Abronia* (Sauria, Anguidae) from the Sierra Madre del Sur of Oaxaca, Mexico. Am. Mus. Novitates 2279: 1-21.

Bostic, D. L.
1971.   Herpetofauna of the Pacific coast of north central Baja California, Mexico, with a description of a new subspecies of *Phyllodactylus xanti*. Trans. San Diego Soc. Nat. Hist. 16: 237-264.

Boulenger, G. A.
1885.   Catalogue of the lizards in the British Museum (Natural History), vol. 2. Taylor and Francis, London. 497 pp.
1913.   Description of new lizards in the collection of the British Museum. Ann. Mag. Nat. Hist., ser. 8, 12: 563-566.

Campbell, J. A.
1982.   A new species of *Abronia* (Sauria, Anguidae) from the Sierra Juarez, Oaxaca, Mexico. Herpetologica 38: 355-361.
1984.   A new species of *Abronia* (Sauria, Anguidae) with comments on the herpetogeography of the highlands of southern Mexico. Herpetologica 40: 373-381.

Cope, E. D.
1864.   Contributions to the herpetology of tropical America. Proc. Acad. Nat. Sci. Phila. 16: 166-181.
1866a. Fourth contribution to the herpetology of tropical America. Proc. Acad. Nat. Sci. Phila. 18: 123-132.
1866b. Fifth contribution to the herpetology of tropical America. Proc. Acad. Nat. Sci. Phila. 18: 317-323.

1868. Sixth contribution to the herpetology of tropical America. Proc. Acad. Nat. Sci. Phila. 20: 305-315.
1869. On the origin of genera. Proc. Acad. Nat. Sci. Phila. 22: 1-80.
1875. Check-list of North American batrachia and reptilia; with a systematic list of higher groups, and an essay on geographical distribution. Bull. U. S. Nat. Mus. 1: 1-104.
1878. Tenth contribution to the herpetology of tropical America. Proc. Amer. Philos. Soc. (1877) 17: 85-98.
1885. Twelfth contribution to the herpetology of tropical America. Proc. Amer. Philos. Soc. 22: 167-194.
1887. Catalogue of the batrachians and reptiles of Central America and Mexico. Bull. U. S. Nat. Mus. 32: 1-98.
1900. The crocodilians, lizards, and snakes of North America. Rept. U. S. Nat. Mus. for 1898: 153-1294.

Criley, B. B.
1968. The cranial osteology of gerrhonotiform lizards. Am. Midl. Nat. 80: 199-219.

Duméril, A. M. C., and G. Bibron
1839. Erpétologie genérale, vol. 5. Roret, Paris, 854 pp

Dunn, E. R.
1936. The amphibians and reptiles of the Mexican expedition of 1934. Proc. Acad. Nat. Sci. Phila. 88: 471-477.

Dunn, E. R., and J. T. Emlen, Jr.
1932. Reptiles and amphibians from Honduras. Proc.Acad. Nat. Sci. Phila. 84: 21-32.

Estes, R.
1963. A new gerrhonotine lizard from the Pliocene of California. Copeia 1963: 676-680.
1983. Handbuch der Paläoherpetologie. Part 10A. Sauria terrestria, Amphisbaenia. Gustav Fischer Verlag, Stuttgart. 249 pp.

Fitch, H. S.
1934. New alligator lizards from the Pacific coast. Copeia 1934: 6-7.
1938. A systematic account of the alligator lizards (*Gerrhonotus*) in the western United States and Lower California. Am. Midl. Nat. 20: 381-424.

Fitzinger, L. I.
1843. Systema reptilium. Braumüller and Seidel, Vindobonae. 106 pp.

Gauthier, J. A.
1982. Fossil xenosaurid and anguid lizards from the early Eocene Wasatch Formation, southeast Wyoming, and a revision of the Anguioidea. Contr. Geol., Univ. Wyoming 21: 7-54.

Gehlbach, F. R., and B. B. Collette
1957. A contribution to the herpetofauna of the highlands of Oaxaca and Puebla, Mexico. Herpetologica 13: 227-232.

Good, D. A.
1987a. An allozyme analysis of anguid subfamilial relationships (Lacertilia: Anguidae) Copeia 1987: 696-701.
1987b. A phylogenetic analysis of cranial osteology in the gerrhonotine lizards (Lacertilia: Anguidae). J. Herp. 21: 283-295.
1988a. Allozyme variation and phylogenetic relationships among the species of *Elgaria* (Squamata: Anguidae). Herpetologica 44: 154-162.
1988b. Allozyme variation and phylogenetic relationships among the species of *Mesaspis* (Squamata: Anguidae). Herpetologica , in press.
1988c. The phylogenetic position of fossils assigned to the Gerrhonotinae (Squamata: Anguidae). J. Vert. Paleo. 8: 188-195.

Good, D. A., and K. Schwenk
1985. A new species of *Abronia* (Lacertilia: Anguidae) from Oaxaca, Mexico. Copeia 1985: 135-141.

Gray, J. E.
1831. A synopsis of the species of the class Reptilia, *in* Griffith edition of Cuvier, Animal kingdom, vol. 9. Whittaker, Treacher, and Co., London. 110 pp.
1838. Catalogue of the slender-tongued saurians, with descriptions of many new genera and species, part 2. Ann. Mag. Nat. Hist., ser. 1, 1: 388-394.
1845. Catalogue of the specimens of lizards in the collection of the British Museum. Edward Newman, London. 289 pp.

Grinnell, J., and H. W. Grinnell
1907. Reptiles of Los Angeles County, California. Throop Inst. Bull. 35: 1-64.

Guillette, L. J., Jr., and H. M. Smith
1982. A review of the Mexican lizard *Barisia imbricata*, and the description of a new subspecies. Trans. Kansas Acad. Sci. 85: 13-33.

Günther, A. C. L. G.
1885. Reptilia and Batrachia. *In* F. D. Godman, and O. Salvin, Biologia Centrali-Americana. R. H. Porter and Dulau, London. 326 pp.

Hallowell, E.
1852. Descriptions of new species of reptiles inhabiting North America. Proc. Acad. Nat. Sci. Phila. 6: 177-182.

Hardy, L. M., and R. W. McDiarmid.
1969. The amphibians and reptiles of Sinaloa, Mexico. Univ. Kansas Publ. Mus. Nat. Hist. 18: 39-252.

Hartweg, N., and J. A. Tihen
1946. Lizards of the genus *Gerrhonotus* from Chiapas, Mexico. Occ. Pap. Mus. Zool., Univ. Michigan 497: 1-16.

Hennig, W.
1966. Phylogenetic Systematics. University of Illinois Press, Urbana. 263 pp.

Hernandez, F.
1651. Rerum medicarum Novae Hispaniae thesaurus. Rome.

Herrick, C. L., J. Terry, and H. N. Herrick
1899. Notes on a collection of lizards from New Mexico. Bull. Sci. Lab. Denison Univ. 11: 117-148.

Hidalgo, H.
1983. Two new species of *Abronia* (Sauria: Anguidae) from the cloud forests of El Salvador. Occ. Pap. Mus. Nat. Hist. Univ. Kansas 105: 1-11.

Holman, J. A.
1973. Reptiles of the Egelhoff local fauna (Upper Miocene) of Nebraska. Contr. Mus. Paleon. Univ. Michigan 24: 125-134.

Karges, J. P., and J. W. Wright
1987. A new species of *Barisia* (Sauria, Anguidae) from Oaxaca, Mexico. Contr. Sci., Nat. Hist. Mus. Los Angeles Co. 381: 1-11.

Knight, R. A., and J. F. Scudday
1985. A new *Gerrhonotus* (Lacertilia: Anguidae) from the Sierra Madre Oriental, Nuevo León, Mexico. Southwest. Nat. 30: 89-94.

Lynch, J. D., and H. M. Smith
1965. New or unusual amphibians and reptiles from Oaxaca, Mexico, I. Herpetologica 21: 168-177.

Maddison, W. P., M. J. Donoghue, and D. R. Maddison
1984. Outgroup analysis and parsimony. Syst. Zool. 33: 83-103.

Martin, P. S.
1955. Herpetological records from the Gomez Farias region of southwestern Tamaulipas, Mexico. Copeia 1955: 173-180.

Martín del Campo, R.
1939. Contribución al conocimiento de los gerrhonoti mexicanos, con la presentación de una nueva forma. An. Inst. Biol. Univ. Mexico 10: 353-361.

McCoy, C. J.
1970. A new alligator lizard (genus *Gerrhonotus*) from the Cuatro Ciénegas Basin, Coahuila, Mexico. Southwest. Nat. 15: 37-44.

McDowell, S. B., and C. M. Bogert
1954. The systematic position of *Lanthanotus* and affinities of the anguinomorphan lizards. Bull. Amer. Mus. Nat. Hist. 105: 1-142.

Meszoely, C. A. M.
1970. North American fossil anguid lizards. Bull. Mus. Comp. Zool. 139: 87-149.

Mocquard, M. F.
1905. Diagnoses de quelques espèces nouvelles de reptiles. Bull. Mus. Natnl. Hist. Nat. Paris 1905: 76-79.

Morafka, D. J.
1977. A biogeographical analysis of the Chihuahuan Desert through its herpetofauna. Dr. W. Junk, B. V., The Hague. 313 pp.

Müller, J. W. von
1865. Reisen in den Vereinigten Staaten, Canada und Mexiko. Beiträge zur Geschichte, Statistik und Zoologie von Mexiko. F. A. Brockhaus, Leipzig. 643 pp.

Murphy, R. W.
1983. Paleobiogeography and genetic differentiation of the Baja California herpetofauna. Occ. Pap. Calif. Acad. Sci. 137: 1-48.

O'Shaughnessy, A. W. E.
1873. Herpetological notes. Ann. Mag. Nat. Hist., ser. 4, 12: 44-48.

Ottley, J. R.
1983. Geographic distribution: *Gerrhonotus paucicarinatus*. SSAR Herp.Review 14: 27.

Peale, T. R., and J. Green
1830. Description of two new species of the Linnaean genus *Lacerta*. J. Acad. Nat. Sci. Phila. 6: 231-234.

Peters, W.
1877. Eine Mittheilung über neue Arten der Sauriergattung *Gerrhonotus*. Monatsb. Akad. Wiss. Berlin 1876: 298-300.

Rieppel, O.
1980. The phylogeny of anguinomorph lizards. Birkhauser Verlag, Basel. 86 pp.

Sanchez-Herrera, O., and W. Lopez-Forment C.
1980. The lizard *Abronia deppei* (Sauria, Anguidae) in the state of México, with the restriction of its type locality. Bull. Maryland Herp. Soc. 16: 83-87.

Savage, J. M.
1960. Evolution of a peninsular herpetofauna. Syst. Zool. 9: 184-212.
1966. The origin and history of the Central American herpetofauna. Copeia 1966: 719-766.
1982. The enigma of the Central American herpetofauna: dispersals or vicariance? Ann. Missouri Bot. Gard. 69: 464-547.

Schmidt, K. P.
1928. Reptiles collected in Salvador for the California Institute of Technology. Field Mus. Nat. Hist., Zool. Ser. 12: 193-201.
1953. A check list of North American amphibians and reptiles. University of Chicago Press. 280 pp.

Skilton, A. J.
1849. Description of two reptiles from Oregon. Amer. J. Sci. Arts 7: 202.

Smith, H. M.
1942. Mexican herpetological miscellany: 3. A tentative arrangement and key to Mexican *Gerrhonotus*, with the description of a new race. Proc. U. S. Nat. Mus. 92: 363-369.
1946. Handbook of lizards. Comstock Publ. Co., Ithaca, N.Y. 557 pp.
1986. The generic allocation of two species of Mexican anguid lizards. Bull. Maryland Herp. Soc. 22: 21-22.

Smith, H. M., and M. Alvarez del Toro
1962. Notulae herpetologicae Chiapasiae III. Herpetologica 18: 101-107.
1963. Notulae herpetologicae Chiapasiae IV. Herpetologica 19: 100-105.

Smith, H. M., and R. B. Smith.
1981. Another epiphytic alligator lizard (*Abronia*) from Mexico. Bull. Maryland Herp. Soc. 17: 51-60.

Smith, H. M., and E. H. Taylor
1950. An annotated checklist and key to the reptiles of Mexico, exclusive of the snakes. Bull. U. S. Nat. Mus. 199: 1-253.

Smith, H. M., M. J. Preston, and R. B. Smith
1983. The history of the concept of viviparity in the alligator lizard genus *Barisia*. Herp. Review 14: 34-35.

Spengler, J. C., and H. M. Smith
1982. A range extension for the alligator lizard *Barisia gadovi levigata*. Bull. Maryland Herp. Soc. 18: 172-174.

Stebbins, R. C.
1958. A new alligator lizard from the Panamint Mountains, Inyo County, California. Am. Mus. Novitates 1883: 1-27.

Stejneger, L.
1890. On the North American lizards of the genus *Barissia* of Gray. Proc. U. S. Nat. Mus. 13: 183-185.
1902. *Gerrhonotus caeruleus* versus *Gerrhonotus burnettii*. Proc. Biol. Soc. Washington 15: 37.
1907. A new gerrhonotine lizard from Costa Rica. Proc. U. S. Nat. Mus. 32: 505-506.

Sumichrast, F.
1882. Enumeración de las especies de reptiles observados en la parte meridional de la Republica Mexicana. La Naturaleza 6: 31-45.

Swofford, D. L.
1985. PAUP, phylogenetic analysis using parsimony, version 2. Illinois Natural History Survey.

Tanner, W. W.
1959. The status of *Gerrhonotus* in Utah. Herpetologica 15: 178-180.

Taylor, E. H., and I. W. Knobloch.
1940. Report on an herpetological collection from the Sierra Madre Mountains of Chihuahua. Proc. Biol. Soc. Washington 53: 125-130.

Tihen, J. A.
1944. A new *Gerrhonotus* from Oaxaca. Copeia 1944: 112-115.

1948. A new *Gerrhonotus* from San Luis Potosí. Trans. Kansas Acad. Sci. 51: 302-305.
1949a. The genera of gerrhonotine lizards. Am. Midl. Nat. 41: 580-601.
1949b. A review of the lizard genus *Barisia*. Univ. Kansas Sci. Bull. 33: 217-256.
1954. Gerrhonotine lizards recently added to the American Museum collection, with further revisions of the genus *Abronia*. Am. Mus. Novitates 1687: 1-26.

Van Denburgh, J.
1897. The reptiles of the Pacific Coast and Great Basin. Occas. Pap. Calif. Acad. Sci. 5: 1-236.
1922. The reptiles of western North America, vol. 1. Lizards. Occas. Pap. Calif. Acad. Sci. 10: 1-611.

Waddick, J. W., and H. M. Smith
1974. The significance of scale characters in evaluation of the lizard genera *Gerrhonotus*, *Elgaria*, and *Barisia*. Great Basin Nat. 34: 257-266.

Wagler, J.
1830. Descriptiones et icones amphibiorum, fasc. 2. J. G. Cotta, Munich.

Webb, R. G.
1962. A new alligator lizard (genus *Gerrhonotus*) from western Mexico. Herpetologica 18: 73-79.

Werler, J. E., and F. A. Shannon
1961. Two new lizards (genera *Abronia* and *Xenosaurus*) from the Los Tuxtlas Range of Veracruz, Mexico. Trans. Kansas Acad. Sci. 64: 123-132.

Wiegmann, A. F.
1828. Beiträge zür Amphibienkunde. Isis von Oken 21: 364-383.
1834. Herpetologia Mexicana, pt. 1, saurorum species. Luderlitz, Berlin. 51 pp.

Wilson, L. D., L. Porras, and J. R. McCranie
1986. Distributional and taxonomic comments on some members of the Honduran Herpetofauna. Milwaukee Publ. Mus. Contr. Biol. Geol. 66: 1-18.

Yanev, K. P.
1980. Biogeography and distribution of three parapatric salamander species in coastal and borderland California. In D. M. Powers (ed.), The California islands: proceedings of a multidisciplinary symposium, Santa Barbara Museum of Natural History, pp. 531-550.

Yarrow, H. C.
1883. Check list of North American Reptilia and Batrachia, with catalogue of specimens in U. S. National Museum. Bull. U. S. Nat. Mus. 24: 1-249.

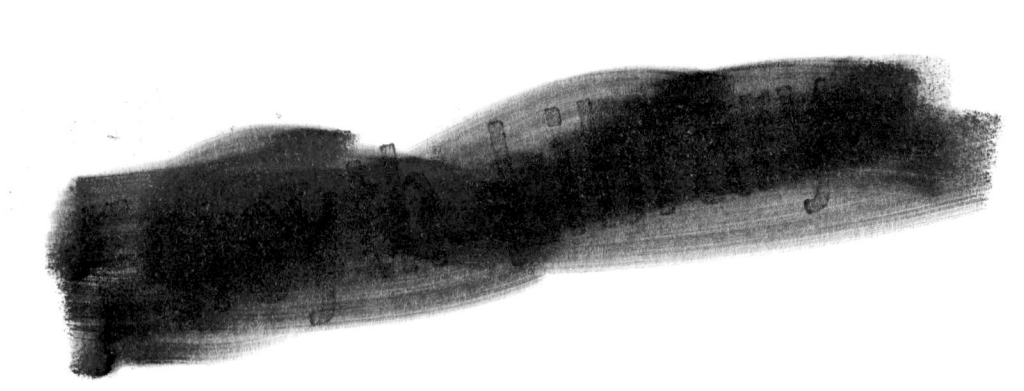

WITHDRAWN